General Chemistry 1 Course Pack

Second Edition

Sapna Gupta
Palm Beach State College

INCLUDES ACCESS TO THE **KHQ** STUDY APP

All chemistry structures from author were created using ACD chemsketch software.
Cover image © Shutterstock.com

Kendall Hunt
publishing company

www.kendallhunt.com
Send all inquiries to:
4050 Westmark Drive
Dubuque, IA 52004-1840

Copyright © 2018 by Kendall Hunt Publishing Company
Copyright © 2020 by Sapna Gupta

Text ISBN: 979-8-7657-2702-7
PAK ISBN: 979-8-7657-2676-1

Kendall Hunt Publishing Company has the exclusive rights to reproduce this work, to prepare derivative works from this work, to publicly distribute this work, to publicly perform this work and to publicly display this work. All rights reserved. No part of this publication may be reproduced, stored in a retrieval system, or transmitted, in any form or by any means, electronic, mechanical, photocopying, recording, or otherwise, without the prior written permission of the copyright owner.

Published in the United States of America

Contents

About the author.. v
About this Course Pack.. vii

Chapter 1   **Chemistry, Matter and Measurement** 1
   Chapter Summary   1
   Power Point: Chemistry: Study of Matter   2
   Power Point: Measurement   9
   Power Point: Temperature and Density   19
   Worksheets   25

Chapter 2   **Atoms, Molecules and Ions** 33
   Chapter Summary   33
   Power Point: Atomic Structure   36
   Power Point: Nomenclature   49
   Worksheets   61

Chapter 3   **Stoichiometry: Chemical Calculations** 69
   Chapter Summary   69
   Power Point: Mass Percent and Empirical Formula   70
   Power Point: Stoichiometry - Equations   83
   Power Point: Stoichiometry - Introduction   89
   Power Point: Stoichiometry – 2 Limiting, Excess Reagent and Percent Yields   95
   Worksheets   101

Chapter 4   **Reactions in Aqueous Solutions** 111
   Chapter Summary   111
   Power Point: Electrolytes and Precipitation Reactions   114
   Power Point: Electrolytes Acid-Base (Neutralization) Oxidation-Reduction
          (Redox) Reactions   125
   Power Point: Solution Concentration   137
   Power Point: Solution Stoichiometry   143
   Worksheets   149

Chapter 5   **Gases** 159
   Chapter Summary   159
   Power Point: Gases -1 Gas Properties and Pressure   160
   Power Point: Gases -2 Gas Laws   165
   Power Point: Gases - 2 Combined Gas Law, Ideal Gas Law and
          Applications of Gas Laws   173
   Power Point: Gases - 4 Gas Stoichiometry   181
   Worksheets   189

Chapter 7   **Quantum Theory** 195
         Chapter Summary   195
         Power Point: Atomic Structure -1 Quantum Model of Atom   197
         Power Point: Atomic Structure -2 Quantum Numbers   207
         Worksheets   215

Chapter 8   **Electronic Configurations, Element Properties and the Periodic Table** 217
         Chapter Summary   217
         Power Point: Electronic Configurations   218
         Power Point: Electronic Configurations – 2 Exceptions, Properties,
                          Ionic Configurations, etc.   227
         Power Point: Periodic Properties   233
         Worksheets   249

Chapter 9   **Chemical Bonds** 255
         Chapter Summary   255
         Power Point: Bonding - 1   256
         Power Point: Bonding – 2 Polar Covalent Bond, Electronegativity,
                          Formal Charge, Resonance   267
         Worksheets   273

Chapter 10   **Chemical Bonding II: Molecular Geometry and Bonding Theory** 279
         Chapter Summary   279
         Power Point: Shapes of Molecules   282
         Power Point: Bonding Theories   293
         Worksheets   307

About the Author

Dr. Sapna Gupta has a Ph.D. in Chemistry and has been teaching chemistry at undergraduate institutions for over 20 years. Her teaching interests are in general, organic, medicinal and forensic chemistry, biochemistry and other non major chemistry courses. She has always tried to learn and incorporate new teaching methods to help her students excel in chemistry. This course pack came as a result of flipping her general chemistry course. In addition to flipping her class, she teaches chemistry online, maintains a website on all her course as well as a YouTube channel for lectures on all her courses. During her spare time, she likes to write children's adventure books and loves to travel (and maintains a blog about it).

About this Course Pack

This course pack is designed for anyone taking a general chemistry 1 course. At Palm Beach State College, it is for CHM1045.

I highly recommend buying the course text book. This course pack can be used in conjunction with, and independently of, any text book. It covers all the learning outcomes of a general chemistry 1 course. (Please note that chapter 6 (Thermochemistry) is missing in this course pack because it will be covered in another course).

Chapter Summary: The chapter summary at the beginning of each chapter has all the formulas, concepts and terminology that is required for that chapter.

PowerPoints: These are the topics explained in detail. There may be 2-4 PowerPoints for each chapter. One chapter has been divided into smaller subtopics so that the material can be learned in small portions, just like in a lecture. The corresponding voice-over of these PowerPoints can be found on my YouTube channel of Dr. Sapna Gupta.

PowerPoint Study Sheets: There is one PowerPoint study sheet corresponding to each PowerPoint lecture. The study sheets should be done concurrently to listening on YouTube or reading the material of that topic. The PowerPoint study sheets are one page of questions that help you learn the material covered in the PowerPoint presentation. The study sheets should be done by each student independently to learn the basic concepts of that chapter.

Worksheets: These are designed to provide deeper understanding of the topic and for the student to be able to solve more complex problems. These will generally be done in class as group work.

It has been observed in educational fields that the more the students work independently on learning the material, the more they understand the material and consequently they perform better in the course.

My hope is that all students who use this course pack will feel that they are better prepared to move on to a higher level chemistry course.

Chapter 1
Chemistry, Matter and Measurement

Chemistry: study of matter
Physical property: e.g. color, solid/liquid/gas
Chemical change: irreversible (rusting, spoiling of milk)

chemical property: e.g. reactivity
physical change (reversible): ice to water

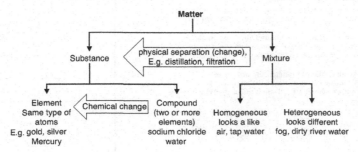

Scientific Method

Observation → Hypothesis → Experimentation (collect data), must be replicable → Theory → Law

Scientific Measurements

Physical Quality	Non SI	SI
Length	Miles, feet	Meters (m)
Weight	Pounds, ounces	Grams (g)
Time	Seconds	Seconds (s)
Temperature	Degree Fahrenheit, Celsius	Kelvin (K)
Electric current	Ampere	Ampere (A)
Volume	Gallons, quarts	Liters (L)
Pressure	Atm, torr, Pascal	Newtons (N)

Giga (G)	10^9
Mega (M)	10^6
Kilo (k)	10^3
Deci (d)	10^{-1}
Centi (c)	10^{-2}
Milli (m)	10^{-3}
Micro (μ)	10^{-6}
Nano (n)	10^{-9}
Pico (p)	10^{-12}

Temperature Unit Conversions: $°C = (°F - 32)/1.8$ AND $°F = 1.8°C + 32$

Measuring Instruments:

<u>Length</u>: meter stick or measuring tape
<u>Solid Volume</u>: meter stick
<u>Temperature</u>: mercury or digital thermometer
<u>Pressure</u>: barometer

<u>Time</u>: stop watch or watch
<u>Liquid volume</u>: measuring cylinder, beakers
<u>Weight</u>: electronic balance, analytic balance

Measurement

<u>Precision</u>: measured values close to each other
<u>Extensive Property</u>: dependent on amount of substance
<u>Intensive Property</u>: independent of amount of substance

<u>Accuracy</u>: measured value close to actual value

Significant figures (SF)

Measured value, depends on the measuring instrument and technique.
E.g. 4.335 – 4 SF   2.09 – 3 SF   200 – 1 SF   0.091 – 2 SF   2.00 – 3 SF

Density (g/mL): d = g/mL (1 cm^3 = 1 mL)

Chapter 1 Chemistry: Study of Matter

Dr. Sapna Gupta

Chemistry

- Study of matter
- Matter is anything that has volume (occupies space) and has mass
- History of Chemistry:
    - Old name: alchemy but later became chemistry
    - Chemistry has always been around whether it was the iron age, bronze age, herbal medicine, discovery of antibiotics, discovery of new elements etc.
    - Chemistry was branched into Organic, Inorganic and Physical initially. Organic has to do with study of chemicals based on carbon, inorganic is everything but carbon, and physical chemistry studies the physical properties of substances.
    - Later chemistry branched out further into analytical (analysis of compounds or mixtures), biochemistry (works on the interface of chemistry and biology), nuclear chemistry (all about radioactivity) etc.
    - Links: Chemical Heritage Foundation

Notes:

Scientific Method

- In order to do good science it is important to follow certain guidelines.
    - **Observation** – observe a phenomenon that needs further study.
    - **Hypothesis** – come up with an idea of what might be happening and what can be done to get results.
    - **Experimentation** – try our your ideas by designing experiments to prove your hypothesis. This is the longest part of scientific method. Experiments have to be verified by peers and be duplicated to make sure they work.
    - **Theory** – once the experiment proves the hypothesis then come up with a theory that explains your observation.
    - **Law** – if the theory stands the test of time then it becomes a law.

Theory and Laws can be challenged and changed in light of new information

Notes:

Chapter 1: Chemistry, Matter and Measurement

Matter

Matter can be classified in three phases.

Gas	Liquid	Solid
Particles are far apart	Particles are closer	Particles are packed
Particles are always in motion	Particles are in motion but slower than gas	Particles are quite static
Gases take the shape of the container	Liquids take the shape of the container	Solids can be molded into shapes

Source: Sapna Gupta

Notes:

Classification of Matter

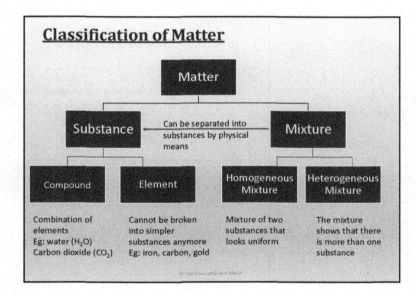

- **Matter**
  - **Substance** (Can be separated into substances by physical means) → **Mixture**
  - **Compound**: Combination of elements. Eg: water (H_2O), Carbon dioxide (CO_2)
  - **Element**: Cannot be broken into simpler substances anymore. Eg: iron, carbon, gold
  - **Homogeneous Mixture**: Mixture of two substances that looks uniform
  - **Heterogeneous Mixture**: The mixture shows that there is more than one substance

Notes:

Chapter 1: Chemistry, Matter and Measurement

Properties of Matter

Physical Property
Study of a property without changing its chemical identity.
E.g.: Physical state, Boiling point, Color

Chemical Property
How chemicals change on reaction with other chemicals.
E.g.: Ability to react with oxygen, Ability to react with fluorine

Changes of Matter

Physical Change
A change that does not change the chemical – and is usually reversible.
E.g.: melting, freezing

Chemical Change
Changes the chemical into something new. Usually reversible only be another chemical change.
E.g.: rusting, burning something in oxygen

Notes:

Properties of Matter

- *Quantitative*: expressed using *numbers*
- *Qualitative*: expressed using *properties*
- *Extensive property*: depends on amount of matter – e.g. mass, length
- *Intensive property*: does not depend on amount – e.g. - density, temperature, color

Notes:

Chapter 1: Chemistry, Matter and Measurement

Key Words/Concepts

- Scientific Method
- Matter
- Hetero and homogeneous mixtures
- Phases of matter: solid, liquid and gases
- Chemical and physical properties
- Chemical and physical change
- Intensive and extensive properties
- Qualitative and quantitative analysis

Notes:

Ch 1/ PowerPoint Study– Study of Matter   Name: _____

Answer these questions as you are watching the videos. They are due in class.
These questions are not just for you to answer but also to prepare you for the exam.
<u>Make sure you understand what you are writing and not just copy from the text book.</u> **Show all work.**

1. <u>Scientific Method</u>:
   a. What are the stages of the scientific method that leads to a law?

   b. Which part of scientific method should be duplicated to be verified?

2. <u>States of Matter</u>: Classify the following into solid gas or liquid.

   Water   Salt   Oxygen   Carbon Dioxide   Gold

3. <u>Mixtures</u>: Circle the homogeneous mixtures in the following.

   Copper penny   Table sugar   Salad Dressing   Dust Storm

<u>Properties of Matter</u>

4. <u>Physical/Chemical</u>: Circle the choice(s) of chemical properties in the properties below.
   a. Density of water is greater than gasoline.
   b. Copper is a great conductor of electricity.
   c. Sodium is changes color in presence of air.

5. <u>Chemical and Physical Change</u>: Circle the choice(s) of chemical changes in the examples below.
   a. Iron rusted in presence of salt and water.
   b. Milk sours to give curdled milk.
   c. Hydrogen conducts electricity.

6. <u>Qualitative/Quantitative</u>: Circle the choice(s) of qualitative analysis below.
   a. The color changes from red to yellow.
   b. The titration requires 2.30 milliliters of acid.
   c. Mixing of two solutions resulted in formation of gas.
   d. The solid formed from the reaction weighed 3.500 grams.

7. <u>Extensive/Intensive</u>: Circle the choice(s) of extensive properties below.
   a. The mass of the water changes as the volume is increased.
   b. Ice melts at 0°C.

Chapter 1   Measurement

Dr. Sapna Gupta

Scientific Method

- Observation
- Hypothesis
- **Experimentation**
- Theory
- Law

Measurement

Any number is useless without any units. Units are used to signify what quantity has been measured.

Physical Quality	Non SI	SI
Length	Miles, feet	Meters (m)
Weight	Pounds, ounces	Grams (g)
Time	Seconds	Seconds (s)
Temperature	Degree Fahrenheit, Celsius	Kelvin (K)
Electric current	Ampere	Ampere (A)
Volume	Gallons, quarts	Liters (L)
Pressure	Atm, torr, Pascal	Newtons (N)

Chapter 1: Chemistry, Matter and Measurement

Instruments Used for Measurement

- **Length**: meter stick or measuring tape
- **Time**: stop watch or watch
- **Solid Volume**: meter stick
- **Liquid Volume**: measuring cylinder, beakers
- **Temperature**: mercury or digital thermometer
- **Weight**: electronic balance, analytical balance
- **Pressure**: barometer

Notes:

Non SI Units

What unit is larger?
Pint or gallon?
   Gallon: 1 gallon = 4 pints
Inch or foot
   Foot: 1 ft = 12 inches
Score or decade?

Which means there are issues with non SI units.

Notes:

SI Units

BaseUnit:
Liter (L)
Meter (m)
Seconds (s)

Prefix	Symbol	Meaning	Example
Tera-	T	1×10^{12} (1,000,000,000,000)	1 teragram (Tg) = 1×10^{12} g
Giga-	G	1×10^{9} (1,000,000,000)	1 gigawatt (GW) = 1×10^{9} W
Mega-	M	1×10^{6} (1,000,000)	1 megahertz (MHz) = 1×10^{6} Hz
Kilo-	k	1×10^{3} (1,000)	1 kilometer (km) = 1×10^{3} m
Deci-	d	1×10^{-1} (0.1)	1 deciliter (dL) = 1×10^{-1} L
Centi-	c	1×10^{-2} (0.01)	1 centimeter (cm) = 1×10^{-2} m
Milli-	m	1×10^{-3} (0.001)	1 millimeter (mm) = 1×10^{-3} m
Micro-	μ	1×10^{-6} (0.000001)	1 microliter (μL) = 1×10^{-6} L
Nano-	n	1×10^{-9} (0.000000001)	1 nanosecond (ns) = 1×10^{-9} s
Pico-	p	1×10^{-12} (0.000000000001)	1 picogram (pg) = 1×10^{-12} g

Notes:

Scientific Notation

- The representation of a number in the form $A \times 10^n$. Where A should be >1 and < 10 and n is an integer
- Every digit included in A is significant.

Example: Write the following numbers in scientific notation:
0.000653
350,000
0.02700

Answer:
6.53×10^{-4}
3.5×10^{5}
2.700×10^{-2}

Notes:

Using Units with Scientific Notations

Example: Write the following measurements without scientific notation using the appropriate SI prefix:

4.851×10^{-9} g = 4.851 ng

3.16×10^{-2} m = 3.16 cm

8.93×10^{-12} s = 8.93 ps

Make the following conversions using scientific notation:

6.20 km to m = 6.20×10^{3} m
2.54 cm to m = 2.54×10^{-2} m
1.98 ns to s = 1.98×10^{-9} s
5.23 µg to g = 5.23×10^{-6} g

Notes:

Measurement

Measurements can be Precise or Accurate
Precision: measured values close to each other
Accuracy: measured value close to actual value

both accurate and precise

not accurate but precise

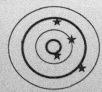
neither accurate nor precise

Source: Sapna Gupta

Notes:

Significant Figures

5.12 g or 5.124 g

Those digits in a measured number (or in the result of a calculation with measured numbers) that include all certain digits plus a final digit having some uncertainty.

5.7 cm
(The tenths place is estimated)

Guidelines for Significant Figures

We need some guideline because in calculations you can have different data of different sig figs;

e.g. in calculation of an area: 1.2356 m x 1.95 m = answer in how many sig figs?

First you need to find how many sig figs you have for each measurement and then do the calculation.

1. Any non-zero digit is significant (1-9)
2. Zeros between non-zero digits are significant (1<u>0</u>1)
3. Zeros to the left of the first non-zero digit are not significant (01<u>0</u>)
4. Zeros to the right of the last non-zero digit are significant if decimal is present (1<u>00.00</u>)
5. Zeros to the right of the last non-zero digit are not significant if decimal is not present (100)

Chapter 1: Chemistry, Matter and Measurement

Solved Problem:
How many significant figures are in each of the following measurements?

a. 310.0 kg = 4 sig figs
b. 0.224800 m = 6 sig figs
c. 0.05930 kg = 4 sig figs
d. 4.380 × 10^{-8} m = 4 sig figs
e. 3.100 s = 4 sig figs
f. 91,000 = 2 sig figs

Notes:

Calculations and Significant Figures

- **Multiplication and Division**: find the lowest sig fig and round off to that number.

    E.g. $\dfrac{6.8914}{1.289 \times 7.28} = 0.734383925 = \underline{0.734}$

- **Addition and Subtraction**: find the lowest decimal number and round off to that number.

    E.g 0.453 − 1.5$\underline{9}$ = -1.13700000 = $\underline{-1.14}$

- **Rounding off**: Find out the correct sig figs you need then, count from the left to that number, look at the number on the immediate right of the last number and if it is 5 or more than 5 then round off the previous number to +1; if lower than 5 then drop that number and all the numbers that follow.

Notes:

Chapter 1: Chemistry, Matter and Measurement

Key Words/Concepts

- Units of Measurement
- SI and non SI units
- Accuracy and Precision
- Significant figures

Notes:

Ch 1/ PowerPoint Study– Measurement

Name: _____

Answer these questions as you are watching the videos. They are due in class.
These questions are not just for you to answer but also to prepare you for the exam. So make sure you understand what you are writing and not just copy from the text book. **Show all work.**

1. <u>Units (SI/non SI)</u>: Write three units of the following units. <u>Write SI unit first</u>.

   Length   Weight   Temperature

2. <u>Scientific Notation (Exponents)</u>: Write the following as scientific notation or the full number.

   3.40×10^4 0.0004758   10589 6.7×10^{-6}

3. <u>Precision and Accuracy</u>: Which of the following measurements, a or b, can be classified as accurate, and why?
   a. A penny was weighed three times and the weight was: 4.50 g, 4.52 g and 4.51 g

   b. Density of water is 1.00 g/mL and in the lab it was found to be: 1.008 g/mL, 1.007 g/mL, 1.008 g/mL

4. <u>Significant Figures</u>: How many significant figures are in the following numbers?

   0.00458 2.98×10^4 107 2.9887×10^{-2} 10.040

5. <u>Rounding off</u>: Round off the following to 3 significant figures. Use exponents for your answer.

   54201   0.0038266   10009

Chapter 1   Temperature and Density

Dr. Sapna Gupta

Temperature Scale

This is a measure of hotness. Heat will flow from higher temperature to lower temperatures.

Units:
- Celsius, °C
- Fahrenheit, °F
- Kelvin, K

Temperature Unit Conversions:
°C = (°F − 32) × 0.56
°F = (1.8 × °C) + 32
K = °C + 273

Notes:

Solved Problem:
In winter, the average low temperature in interior Alaska is −30.°F (two significant figures). What is this temperature in degrees Celsius and in kelvins?

$t_C = (t_F - 32°F)\dfrac{5°C}{9°F}$

$t_C = (-30.°F - 32°F)\dfrac{5°C}{9°F}$

$t_C = (-62°F)\dfrac{5°C}{9°F}$

$t_C = -34.4444444°C$

$t_C = -34°C$

$t_K = \left(t_C \times \dfrac{1K}{1°C}\right) + 273.15\,K$

$t_K = \left(-34°C \times \dfrac{1K}{1°C}\right) + 273.15\,K$

$t_K = -34\,K + 273.15\,K$

$t_K = 239.15\,K$

$t_K = 239\,K$

Notes:

Chapter 1: Chemistry, Matter and Measurement

Derived Units

- These are a combination of the same unit (m^2) or two different units (m/s).

Quantity	Definition of Quantity	SI Unit
Area	length × length	m^2
Volume	length × length × length	m^3
Density	mass per unit volume	kg/m^3
Speed	distance per unit time	m/s
Acceleration	change in speed per unit time	m/s^2

Notes:

Density

- Mass per unit volume
- Units: g/cm^3 (solids), g/mL (liquids and gases)

Solved Problem:

Oil of wintergreen is a colorless liquid used as a flavoring. A 28.1-g sample of oil of wintergreen has a volume of 23.7 mL. What is the density of oil of wintergreen?

m = 28.1 g
V = 23.7 mL

$$d = \frac{m}{V}$$

$$d = \frac{28.1 \text{ g}}{23.7 \text{ mL}}$$

$$d = 1.18565491 \frac{g}{mL}$$

$$d = 1.19 \frac{g}{mL}$$

Notes:

Solved Problem:
A sample of gasoline has a density of 0.718 g/mL. What is the volume of 454 g of gasoline?

$$m = 454 \text{ g}$$
$$d = 0.718 \text{ } \frac{g}{mL}$$

$$d = \frac{m}{V} \qquad V = \frac{m}{d}$$

$$V = \frac{454 \text{ g}}{0.718 \text{ } \frac{g}{mL}}$$

$$V = 632.311978 \text{ mL}$$

$$V = 632 \text{ mL}$$

Notes:

Dimensional Analysis

- A systematic way of calculating by using units during calculations.
- Start with what you know and then sequentially use conversion factors to get the right answer.
- Tips for Problem Solving
  - Read carefully; find information given and what is asked for
  - Find appropriate equations, constants, conversion factors
  - Check for sign, units and significant figures
  - Check for reasonable answer

Notes:

Solved Problem:
Convert 12.00 inches to meters.
Conversion factors needed:
  2.54 cm = 1 in and 100 cm = 1 meter

$$12.00 \, \text{in} \times \frac{2.54 \, \text{cm}}{1 \, \text{in}} \times \frac{1 \, \text{m}}{100 \, \text{cm}} = 0.3048 \, \text{m}$$

Solved Problem:
The Food and Drug Administration (FDA) recommends that dietary sodium intake be no more than 2400 mg per day. What is this mass in pounds (lb), if 1 lb = 453.6 g?

$$2400 \, \text{mg} \times \frac{1 \, \text{g}}{1000 \, \text{mg}} \times \frac{1 \, \text{lb}}{453.6 \, \text{g}} = 5.3 \times 10^{-3} \, \text{lb}$$

Notes:

Key Words/Concepts

- Temperature
- Density
- Dimensional Analysis

Notes:

Ch 1/ PowerPoint Study– Temperature-Density   Name: _____

Answer these questions as you are watching the videos. They are due in class.
These questions are not just for you to answer but also to prepare you for the exam. So make sure you understand what you are writing and not just copy from the text book. **Show all work.**

1. <u>Temperature</u>: (use the formula given. You can skip the fractions: 5/9=0.56 and 9/5=1.8). *Remember:* To convert F to K, you have to convert F to °C first.
   a. Convert 32 °C to F
   b. Convert 46.9 °F to K

2. <u>Unit Conversion</u>: Carry out the following unit conversions.
   a. 20 mL to L

   b. 3.09 kg to g

   c. 3.00 cm to in (1 in = 2.54 cm)

   d. 4.90 oz to mL (1 L = 33.8 oz) (Hint: convert mL to L first)

3. Density: (d=m/v)
   a. Calculate the density of a solid that has a mass of 2.980 g and volume of 8.20 mL.

   b. What is the volume of a 20.9 g solid with a density of 6.90 g/cm^3? (Ans: 3.03 mL)

   c. What is the weight of 23 mL of gasoline which has a density of 0.87 g/mL? (Ans: 20 g)

Ch 1/ Worksheet/Matter   Name: _____

1. In the following, label the things that are matter:

   Iron   philosophy   exhaust gas   human body   an idea.

2. Label the following as chemical or physical properties:
   a. sulfur is yellow   b. natural gas burns,

   c. diamond is hard;   d. iron melts at 1535 °C.

3. Label the following as chemical or physical change:
   a. wool is spun into yarn;   b. cake is baked from flour, eggs, sugar, oil;

   c. milk turns sour;   d. cutting a lawn;

   e. a piece of sodium is cut into small pieces.

4. Which of the following are homogenous or heterogenous mixtures:
   a. gasoline;   b. raisin pudding;   c. vanilla ice cream;

   d. water from the tap;   e. Italian salad dressing.

5. Which of the following are elements and which are compounds?

   Carbon   chlorine   table salt   carbon dioxide   iron.

Ch 1/ Worksheet/Significant Figures/ Rounding Off/Exponents

Name: _____

1. Write the following in exponent form and rounding off to 3 significant figures.

10,000	50,000	0.000004
0.0001	2,000,000,000	0.6200
10,000,000,000	0.000197	328,500
76,450	0.9410	3005

2. Write the following numbers in the "long form":

3.2×10^{-2}	14.3×10^2	4.3×10^3
6.854×10^{14}	9.065×10^{-4}	5×10^{-6}

3. Carry out the following calculations:

Addition and subtraction	Multiplication Division
a. $3.461728 + 14.91 + 0.980001 + 5.2631 =$	a. $6305/0.010 =$
b. $23.1 + 4.77 + 125.39 + 3.581 =$	b. $12.5 \times 75 =$
c. $22.101 - 0.9307 =$	c. $(6.78 \times 10^{-4}) \times (1.4 \times 10^2) =$
d. $0.04216 - 0.0004134 =$	d. $(6.432 + 83)/2.143 =$
e. $564,321 - 264,321 =$	e. $(6.432 + 83.42)/2.14 =$
	f. $14.3 + (12.2 \times 2.0) =$

Ch 1/Worksheet/Unit Conversions

Name: _____

1. Carry out the following conversions:
   a. How many cm in 18.9 inches? (1 in = 2.54 cm)

   b. How many micrograms in 6.8×10^{-7} ounces? (1 oz = 28.35 g)

   c. How many mL are in 0.037 quarts? (1 qt = 0.946 L)

   d. How many grams are in 0.397 pounds (lbs)? (1 kg = 2.21 lb)

   e. 22.6 m to miles (1 mi = 1.609 km)

   f. 289 mL to L

   g. convert 4.5 m into ft (1 ft = 30 cm)

   h. 10.6 cm^3 to m^3

2. How much will milk in Canada cost (in Canadian $/L) if it costs $3.57/gallon in Palm Beach (1 US dollar = 1.01 Canadian dollar and 1 gallon = 3.78 L)

Ch 1/Worksheet/Density

Name: _____

Show all your work in the following calculations and calculate to the correct number of significant figures and write the units.

1. What is the density of a salt solution if 50.0 mL has a mass of 57.0g? (Ans: 1.14 g/mL)

2. Some metal chips have a volume of 3.29 cm^3. The mass of the chips is 16.293 g. What is the density of the metal? (Ans: 4.95 g/cm^3)

3. A glass container weighs 48.462 g. A sample of 4.00 mL of antifreeze is added to the container, the new weight is 54.513 g. What is the density of the antifreeze? (Ans: 1.51 g/mL)

4. Silica has a density of 2.20 g/cm^3. Calculated the volume of silica in 125 Kg. (Ans: 5.68 x 10^4 cm^3)

5. Hylon is a starch bases substance to make polymers. It has a bulk density of 12.8 kg/m^3. How many grams of the material are needed to fill a volume of 2.00 ft^3? (Ans: 725 g)

Chapter 2
Atoms, Molecules and Ions

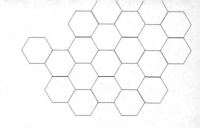

History

John Daltons Atomic Theory:
1. all matter is composed of small indivisible particles called atoms
2. all atoms in an element are same but different elements have different atoms
3. compounds are formed when different atoms combine in fixed proportion
4. in a chemical reaction, atoms are rearranged, no new atoms are created and none are destroyed.

History of Discovery of the Atom: Classical View

1895	Wilhelm Roentgen	Discovers X-rays
1896	Henri Becquerel	Discovers radioactivity
1897	J.J. Thomson	Measured deflection of cathode rays in magnetic and electric fields. Shows that deflections were same regardless of gas in cathode tube. Hence: Rays were not atoms but charged particle found in all matter. Measured the mass to charge ratio: found it to be -5.686×10^{-12} kg/Coulomb, thus smaller than a hydrogen ion (as known previously)
1909	Robert Milliken	Oil-drop experiment to determine the charge on electron. -1.602×10^{-19} C
Early 1900s	Ernest Rutherford	Develops the basic atomic model, establishing that there is a positive center (protons) to the atom which is surrounded by negative charge (electrons).
1932	James Chadwick	Discovers neutrons (have no charge and have same mass as protons).

Basic Structure

Central nucleus surrounded by electrons. Nucleus has protons and neutrons,

Particle	Found in	Charge	Mass
Neutron (n)	Nucleus	None	1 amu
Proton (p)	Nucleus	+	1 amu
Electron (e)	Shells outside nucleus	-	0.000545 amu 9.109×10^{-28} g

Every neutral atom has same number of electrons and protons.
Mass number (M) = n + p
Atomic number (Z) = p (or e)

Isotopes: different mass number but same atomic number.
Allotropes: two or more forms of the same element.

Chapter 2: Atoms, Molecules and Ions

History of the Periodic Table (PT)

1830 – 55 elements were known.
1869 – Dimitri Mendeleev, constructed the first what is now called Periodic Table
Initially PT was arranged by mass number. Most elements that came in a column had similar properties, however some did not. He moved those elements out of sequence to fit the trend. Then PT was arranged according to atomic number.
Modern PT – Glenn Seaborg improved the old PT after he discovered majority of the lanthanide elements.

Molecules and Ionic Compounds
Empirical formula – lowest ratio of elements in a compound.
Molecular formula – actual ratio of elements in a compound.
Binary compound – has two different elements in a compound.
Atomic ions – atoms that have lost or gained electrons
Polyatomic ions – are also compounds that are electrically positive or negative.
Ionic compounds – have anions (-) and cations (+)

Naming
Molecules:

1. Name first element first followed by second element and end the second element name with "ide:
   e.g. HCl - hydrogen chloride
   NO – nitrogen oxide
   For polyatomic molecules:
   N_2O_4 – dinitrogen tetraoxide

2. Most common non-metal "ides" – carbide, nitride, oxide, phosphide, sulfide, fluoride, chloride, bromide, iodide.

3. Prefixes –
   1 –mono; 2- di; 3- tri; 4- tetra; 5- penta; 6- hexa; 7- hepta; 8- octa; 9- nona; 10 – deca.

4. Always write the number of atoms in a molecule as subscripts.

Ionic Compounds:
Made from cations and anions.
<u>Cations</u> – positively charged, formed from metals.
Transition metals can have more than one kind of cations. Have to write 2+ as roman numerals (II) or 3+ as (III) i.e. use roman numerals to show the charge of transition metals. Roman numerals have to be in parenthesis. This is ONLY for transition metals.
<u>Anions</u> – negatively charged, formed from non-metals.

Name metal as is and then the non-metal as "ide" (as in molecules above). Main difference here is – you CANNOT use mono, di, tri etc…as for molecules.
NaCl – sodium chloride
$MgCl_2$ – magnesium chloride (not magnesium dichloride!)

Polyatomic Ions:
<u>Polyatomic Ions</u>: two or more non metals combine to form an ion.
<u>Positive Polyatomic Cations</u>
H_3O^+ hydronium ion (exists only in acidic solutions)
NH_4^+ ammonium ion (formed from NH_3 – ammonia)
<u>Simple Polyatomic Anions</u>
OH^- hydroxide
CN^- cyanide
<u>Oxygen Containing Polyatomic Ions (all end in "ate" or "ite")</u>

Formula	Name
CO_3^{2-}	Carbonate
HCO_3^-	Hydrogen carbonate
SO_4^{2-}	Sulfate
HSO_4^-	Hydrogen sulfate
SO_3^{2-}	Sulfite
HSO_3^-	Hydrogen sulfite
NO_3^-	Nitrate
NO_2^-	Nitrite
PO_4^{3-}	Phosphate
HPO_4^{2-}	Hydrogen phosphate
$H_2PO_4^-$	Dihydrogen phosphate
$C_2H_3O_2^-$ CH_3COO^-	Acetate
CrO_4^{2-}	Chromate
$Cr_2O_7^{2-}$	Dichromate
MnO_4^-	Permanganate
SCN^-	Thiocyanate
$S_2O_3^{2-}$	Thiosulfate
ClO^-	Hypochlorite
ClO_2^-	Chlorite
ClO_3^-	Chlorate
ClO_4^-	Perchlorate

Acids:
Acids: produce hydrogen ions and usually end in "ic" in their names. (Web List)
Hydrates: association of ionic compounds with water molecules. Use mono, di, tri etc. to indicate number of water molecules.

Chapter 2 Atomic Structure

Dr. Sapna Gupta

Atomic Theory

- The first proposal of atom came from a Greek philosopher, Democritus during 5th century. He proposed that atoms were indivisible particles making up everything.
- In 1808 John Dalton, an English scientist studied this further.

© Georgios Kollidas/Shutterstock.com

Notes:

Dalton's Atomic Theory

1. All matter is composed of indivisible **atoms**.
2. An **atom** is an extremely small particle of matter that retains its identity during chemical reactions.
3. An **element** is a type of matter composed of only one kind of atom.
4. A **compound** is composed of atoms of two or more elements chemically combined in fixed proportions.
5. A **chemical reaction** involves the rearrangement of the atoms present in the reacting substances to give new chemical combinations. So no atom is created nor destroyed. (also the *Law of Conservation of Matter*)

Example: Combination of oxygen and carbon to form carbon dioxide.

$$O_2 + C \longrightarrow CO_2$$

Notes:

Law of Definite Proportions

This law states that different samples of a given compound always contain the same elements in the same mass *ratio*.

Sample	Mass of O (g)	Mass of C (g)	Ratio (g O: g C)
124 g carbon dioxide	89.3	33.5	2.66:1
50.4 g carbon dioxide	36.6	123.8	2.66:1
88.6 g carbon dioxide	64.3	24.1	2.66:1

Sample	Mass of O (g)	Mass of C (g)	Ratio (g O: g C)
16.3 g carbon monoxide	9.31	6.98	1.33:1
50.4 g carbon monoxide	14.7	11.1	1.33:1
88.6 g carbon monoxide	50.4	37.8	1.33:1

Notes:

Law of Multiple Proportions

This law states that if two elements can combine to form more than one compound with each other, the masses of one element that combine with a fixed mass of the other element are in ratios of small whole numbers.

$$\frac{\text{ratio of O to C in carbon dioxide}}{\text{ratio of O to C in carbon monoxide}} = \frac{2.66}{1.33} = 2:1$$

Notes:

Atomic Structure

- There are three particles in the atom. Electron, proton and neutron. Neutrons and protons are found in the nucleus (center of the atom) and electrons are outside. Below are the properties of the three particles.
- There are a number of scientists involved and it took more than 30 years to come up with the structure of the atom, and its still not quite done!

Particle	Found in	Charge	Mass
Neutron (n)	Nucleus	None	1 amu 1.673×10^{-24} g
Proton (p)	Nucleus	+	1 amu 1.673×10^{-24} g
Electron (e)	Outside nucleus	−	0.000545 amu 9.109×10^{-28} g

Discovery of the Electron

- Electron was discovered by J.J. Thompson in 1897.
- He used the cathode ray tube and observed that the "ray" coming out of the hydrogen gas was deflected by a magnetic plate.
- He concluded that the particles making up the cathode ray are negative and no matter of the element he used he got the same results.
- He named these particles electron.
- He then calculated the mass to charge ratio.
  (-1.76×10^8 C/g; C = *coulomb*)
- Next slide has the cathode ray experiment.

Chapter 2: Atoms, Molecules and Ions

Cathode Ray Tube Experiment

- See this video to get an idea of how this experiment works.
  http://youtu.be/O9Goyscbazk

Notes:

Robert Milliken's Experiment

- Robert Milliken established the charge of the electron from his "oil drop experiment". –1.66022 x 10^{-19} C
- This helps to establish the mass of the electrons using Thompson's mass/charge ratio.

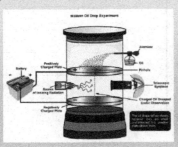

Notes:

Rutherford's Gold Foil Experiment

- If there was a negative particle there has to be a positive – the location of this positive particle (proton) was not known.
- Rutherford helped to establish the location of protons in the atom using his gold foil experiment.
- Watch this video on youtube to get an idea of this experiment. http://youtu.be/5pZj0u_XMbc
- But there was one problem, if Hydrogen has a mass of 1 and has 1 proton, then Helium should be a mass of 2 because it has two protons; however Helium has a mass of 4.
- So there must be another particle.
- James Chadwick in 1932 discovered the last particle, neutron which is also present in the nucleus. It was the last to be discovered because it is neutral.

Notes:

Rutherford's Gold Foil Experiment

On radiating the atom with a positive radiation, he expected radiation to pass through the atom at all locations. He expected that the positive was distributed all over the atom (plum pudding model). However this was not observed

The gold foil emitted radiation at specific locations and some radiation simply bounced back which was completely unexpected.

He concluded that the positive protons were located as one big mass in the center of atom rather that distributed all over the atom. This is the nuclear model of atom.

Expected result

Rutherford's Gold Foil Experiment
- Most α Particles Travel through the Foil Undeflected
- Some α Particles are Deflected by Small Angles
- Few α Particles Travels Back from the Foil
- Gold Foil
- Detector
- Beam of α Particles
- Radioactive Source
- Lead Shield

Actual result

© udaix/Shutterstock.com

Notes:

Chapter 2: Atoms, Molecules and Ions

Atomic Structure and Properties

- Atom has electrons, protons and neutrons.
- Neutrons and protons are in the nucleus whereas electrons are outside.
- Mass of the atom is primarily from the nucleus, i.e. neutrons and protons. Electrons are not measured in the mass of the atom.
- Number of protons and electrons in an electrically neutral atom are equal.

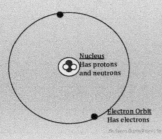

Source: Sapna Gupta

Notes:

Atomic Number and Mass Number

- **Atomic number (Z)** is the number of protons
- **Mass number (A)** is the number of protons and neutrons of an atom.
- Identity of atom is by the number of protons.

Mass number (number of protons + neutrons)

Atomic number (number of protons)

$${}^{A}_{Z}X$$ Atom Symbol

- In the periodic table the atomic number is at the top because the periodic table is arranged chronologically by the atomic number.
- Look at the periodic table – the mass should be a whole number if mass is the addition of two types of particles – but its not; so there is something else going on.
- **Isotopes are elements with the same atomic number but different mass number.**
- Each element has some isotopes. One cannot predict how many isotopes or what the mass will be. This has to be done experimentally.

Notes:

Isotopes

Isotopes are detected by Mass Spectrophotometer. This instrument measures the percent of different atoms in one element. (*Note: this challenges Daltons atomic theory as he said that all atoms are the same in an element. Well...they still are. The number of protons in an element don't change, only neutrons change which does not change the identity of the element.*)

The output looks like below where one can see the abundance of each isotope and its mass.

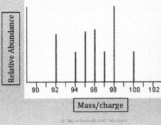

Source: Sapna Gupta

Calculation of Atomic Mass

For example lead has four naturally occurring isotopes. The mass and percentage of each isotope are as found as below. One has to calculate the average mass of the naturally occurring element.

Percentage Abundance	Mass (amu)
1.48	203.973
23.6	205.9745
22.6	206.9759
52.3	207.9766

Fractional Abundance	Mass (amu)	Mass From Isotope
0.0148	203.973	3.01$_{880040}$
0.236	205.9745	48.6$_{099820}$
0.226	206.9759	46.7$_{765534}$
0.523	207.9766	108.$_{771762}$
Addition of all fractions of mass contribution of isotopes gives the actual mass		207.$_{177096}$

Chapter 2: Atoms, Molecules and Ions

The Periodic Table and Elements

- Elements are arranged in a specific order in groups and rows.
- The periodic table was developed first by Dimitri Mendeleev.
- See a short video here: http://youtu.be/-kUg_KJhcLo
- Here is the best periodic table with videos and history of elements. http://www.rsc.org/periodic-table

The First Periodic Table

Air
Water
Fire
Earth

Source: Sapna Gupta

Notes:

Division of the Periodic Table

- The periodic table is very thorough in its arrangement and can give a lot of information.
- The PT is arranged chronologically according to atomic number.
- The PT is divided into metals, metalloids and non metals.
- Left PT – all metals, right PT – all nonmetals and in the middle are metalloids.
- Hydrogen is a non metal even though its on the left side.

Metals	Nonmetals	Metalloids
•Shiny, lustrous, malleable	•Vary in color, not shiny	•Can behave like a metal or non metal depending on the conditons.
•Good conductors of heat and electricity	•Poor conductors	
•Hard	•Brittle	
•Give electrons to form cations	•Accept electrons to form anions	
	•Group VII and VIII are all non metals	

Source: Sapna Gupta

Notes:

Names of Groups in the Periodic Table

- Some groups have names, some don't; they are referred to as group numbers (III-VI).
- Groups I-VIII labeled below are called the main group elements.
- Hydrogen is not an alkali metal even though its in Group 1

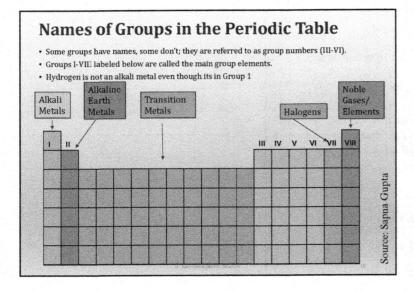

Source: Sapna Gupta

Elements and Symbols

- All elements are represented by symbols of one or two alphabets.
- Symbols are very useful when writing chemical formulas and equations.
- The appropriate way of writing symbols is first letter capital and second small case. There is NO other way.
- Some elements are easy to know e.g. H for Hydrogen and He for Helium while some are harder e.g. Na for Sodium and K for Potassium. That could be because the origin of the name is either Greek, Latin or another language.
- Names of elements are based on planets, people, country etc.
- Go the website (*given earlier also*) http://www.rsc.org/periodic-table to learn about the discovery, names etc. of the elements.
- As a chemistry student you must know elements 1-20 by heart. Other key elements are: all of group I, II, VII, VIII, transition metal row 1. The rest will be mentioned in class.

Solved Problems

1) Give the number of protons (p), electrons (e), and neutrons (n) in one atom of chlorine-37.
a) 37 p, 37 e, 17 n
b) 17 p, 17 e, 37 n
c) 17 p, 17 e, 20 n
d) 37 p, 17 e, 20 n
e) 17 p, 37 e, 17 n

2) An aluminum ion, Al^{3+}, has:
a) 13 protons and 13 electrons
b) 27 protons and 24 electrons
c) 16 protons and 13 electrons
d) 13 protons and 10 electrons
e) 10 protons and 13 electrons

3) Which of these elements is most likely to be a good conductor of electricity?
a) N
b) S
c) He
d) Cl
e) Fe

4) Who is credited with discovering the atomic nucleus?
a) Dalton
b) Gay-Lussac
c) Thomson
d) Millikan
e) Rutherford

Key Words and Concepts

- Daltons Atomic Theory
- Law of conservation of matter
- JJ Thompson
- Milliken's oil drop experiment
- Rutherford's gold foil experiment
- Electron, Proton, neutron
- Nuclear model of the atom
- Atomic number
- Atomic mass
- Isotopes
- Periodic table
- Metals, metalloids and non metals
- Alkali metal, alkaline earth metals, halogens, noble gases, transition metals

Ch 2/ PowerPoint Study–02-1 Atomic Theory of Matter   Name: _____

Answer these questions as you are watching the videos. They are due in class.
These questions are not just for you to answer but also to prepare you for the exam.
Make sure you understand what you are writing and not just copy from the text book. **Show all work.**

1. Which law is supported by each of the following statements? (The laws are: conservation of mass, definite proportion, and multiple proportions.)
    a. In hydrogen peroxide there are 15.9 grams of oxygen per 1.00 g of hydrogen and in water there are 7.94 grams of oxygen per 1.00 g of hydrogen.

    b. The total mass of reactants (starting materials) is the same as the total mass of products when a chemical reaction is carried out in a closed system.

    c. In any sample of a given compound, the mass proportion of each element is the same.

2. Who was the scientist to discover the electron? _____

3. Which scientist calculated the electronic charge? _____

4. Briefly explain the gold foil experiment? What was the purpose of doing this experiment?

5. What are the three subatomic particles and the charges on the particles found in an atom?

6. Which two atomic particles are equal in number? _____

7. What is an isotope?

8. Fill in the following table with the appropriate name and/or symbol of the element. Classify them in their group, if applicable.

Neon	Ca	P	Iron
Ti	Tin	I	Zinc

9. Write the number of all the three particles in the following table. Remember to round off to whole numbers. Number of particles are always in whole numbers, never in fractions.

Element	Electrons	Protons	Neutrons
Silicon			
Sulfur – 31 (atomic mass 31)			
Ca²⁺ Calcium lost 2e⁻			
N³⁻ Nitrogen gained 3e⁻			

Chapter 2 Nomenclature

Dr. Sapna Gupta

Introduction

- Compounds are elements combined in different proportions.
- There are two main types of combinations (bonding) as shown below.
- Nomenclature is of three kinds – two different types for ionic – one for main group and one for transition metals; and one type for covalent.

Bonding

Ionic
- Transfer of electrons
- Between metals and non metals
- Cations and Anions

Covalent
- Sharing of electrons
- Between non metals only
- No ions

Molecules and Formula Units

- Covalent compounds are called molecules.
- Formula unit is the smallest ratio of cations to anions in an ionic compound. In solid state ionic compounds a specific number of ions are associated with each other. These form the crystal structure.
- There are seven diatomic gases/molecules whose names are the same as the element names:
  $H_2, O_2, N_2, F_2, Cl_2, Br_2$ and I_2.

Chapter 2: Atoms, Molecules and Ions

Representing Combinations

- **Empirical Formula** – lowest ratio of combination of elements e.g. H_2O, CH_2O.
- **Molecular Formula** – actual ratio of elements in a compound. For ionic compounds the empirical formula is also the molecular formula (formula unit). For covalent compounds sometimes the empirical formula can be multiplied by an integer to give the molecular formula. E.g. $(CH_2O) \times 6 = C_6H_{12}O_6$ (glucose)
- **Structural Formula** – 3D representation of the ionic or covalent compound.

- Example: H_2O_2 is molecular formula, HO is the empirical formula of that compound.

Notes:

Ionic Compounds

- Formed from metals and non metals.
- Metals lose electrons to form cations.
- Non metals gain electrons to form anions.
- Electrons transfer from metals to non metals.
- Final unit is electrically neutral.

Metals lose e⁻	Cation formed E.g.	Non Metals gain e⁻	Anion formed E.g.
Group I 1e⁻ lost	Li^+, Na^+	Group V 3e⁻ gain	N^{3-}, P^{3-}
Group II 2e⁻ lost	Mg^{2+}, Ca^{2+}	Group VI 2e⁻ gain	O^{2-}, S^{2-}
Group III 3e⁻ lost	Al^{3+}	Group VII 1e⁻ gain	Cl^-, Br^-

Notes:

Naming Ionic Compounds – Main Group

- Naming cations
  - Name the element and add the word "ion"
  - Example: Na^+, sodium ion
- Naming Anions
  - Name the element and modify the ending to "-ide"
  - Example: Cl^- (chloride), O^{2-} (oxide)
- <u>Naming ionic compounds</u>: leave the name of the metal as is, change the name of the non metal to element ending with **–ide**. For example, sodium chloride, calcium oxide.
- <u>Note</u>: The name does not say anything about the ratio of the elements. This is because there is only one combination possible for main group elements.
- There are two exceptions –
  Pb in Group IV has two ions Pb^{2+} and Pb^{4+}; and
  Tl in Group III has two ions Tl^+ and Tl^{3+}

Forming Compounds (Formula Units)

Compound formed is electrically neutral so the sum of the charges on the cation(s) and anion(s) in each formula unit must be zero. The cris-cross method is given below to easily form compounds.

Al^{3+} and O^{2-} Al_2O_3

Naming Ionic Compounds – Transition Metals

- Cations from transition metals with some exceptions
  - Name element
  - Indicate charge of metal with Roman numeral
  - Add word "ion"
  - Example: Cu^{2+}, copper(II) ion, Fe^{3+}, Iron (III) ion

Example: form the compounds between the following ions.

Fe^{3+} and O^{2-} → Fe_2O_3
Iron (III) oxide

Mn^{4+} and S^{2-} → $Mn_2S_4 = MnS_2$
Manganese (IV) sulfide

More on Ions

- Main group elements give and take definite number of electrons.
- Transition metals however give a different number of electrons. There is no point memorizing the charges; it is better to look at the formula unit or name to figure out the charge (valency) on the metal.
- The table below gives the cations and anions you need from PT.

Source: Sapna Gupta

Polyatomic Ions

Positive Polyatomic Cations
H_3O^+ hydronium ion (exists only in acidic solutions)
NH_4^+ ammonium ion (formed from NH_3: ammonia)

Simple Polyatomic Anions
OH^- hydroxide
CN^- cyanide

Polyatomic Ions Containing Oxygen
(all end in "ate" or "ite")

- These are ions (cations and anions) formed from combination of non metals.
- Most polyatomic ions are anions meaning they have excess electrons. (*Where are these excess electrons coming from?* From the metals they combine with).

Formula	Name
CO_3^{2-}	Carbonate
HCO_3^-	Hydrogen carbonate
SO_4^{2-}	Sulfate
HSO_4^-	Hydrogen sulfate
SO_3^{2-}	Sulfite
HSO_3^-	Hydrogen sulfite
NO_3^-	Nitrate
NO_2^-	Nitrite
PO_4^{3-}	Phosphate
HPO_4^{2-}	Hydrogen phosphate
$H_2PO_4^-$	Dihydrogen phosphate
$C_2H_3O_2^-$ CH_3COO^-	Acetate

Formula	Name
CrO_4^{2-}	Chromate
$Cr_2O_7^{2-}$	Dichromate
MnO_4^-	Permanganate
SCN^-	Thiocyanate
$S_2O_3^{2-}$	Thiosulfate
ClO^-	Hypochlorite
ClO_2^-	Chlorite
ClO_3^-	Chlorate
ClO_4^-	Perchlorate

Notes:

Compounds Containing Polyatomic Ions

Polyatomic ions form compounds with cations or anions as usual.

Example:
- Na^+ and OH^- is NaOH – sodium hydroxide
- K^+ and SO_4^{2-} is K_2SO_4 – potassium sulfate
- Ca^{2+} and PO_4^{3-} is $Ca_3(PO_4)_2$ – calcium phosphate
- Cr^{2+} and HCO_3^- is $Cr(HCO_3)_2$ – chromium (II) hydrogen carbonate
- NH_4^+ and NO_3^- is NH_4NO_3 – ammonium nitrate

Notes:

Naming Covalent Compounds

- Combination of non metals gives covalent compounds
- The first element is named as such and the second one ends with "ide"
- In a number of cases more than one combination is possible for covalent compounds hence "mono", "di" etc are used to indicate how many atoms are in the compound.
- Note: if the first element is only one then don't indicate it as "mono"

Number	Prefix
1	mono
2	di
3	tri
4	tetra
5	penta
6	hexa
7	hepta
8	octa
9	nona
10	deca

Examples
Oxygen difluoride - OF_2
Tetrasulfur tetranitride - S_4N_4
Boron trichloride - BCl_3
Carbon disulfide - CS_2
Nitrogen tribromide - NBr_3
Dinitrogen tetrafluoride - N_2F_4

Notes:

Acids and Bases

Acids	Bases
1. Give protons 2. Corrosive 3. Sour 4. Found in fruit juices 5. Reacts with metals to give H_2 gas	1. Accept protons from acids 2. Caustic 3. Bitter 4. Found in cleaners
Examples: Mineral acids/Inorganic acids Strong acids - Sulfuric acid, nitric acid, hydrochloric acid Weak acids – phosphoric acid, acetic acid, carbonic acid	Examples: Inorganic bases Strong bases – sodium hydroxide, potassium hydroxide Weak bases – ammonium hydroxide

Notes:

Naming Acids

There are two kinds of acids – binary and oxo-acids

Binary acids (Haloacids) Made from hydrogen halide dissolved in water		Oxoacids Made from polyatomic anions.	
HF	Hydrofluoric acid	HNO_3	Nitric acid
HCl	Hydrochloric acid	H_2SO_4	Sulfuric acid
HBr	Hydrobromic acid	H_3PO_4	Phosphoric acid
HI	Hydroiodic acid	H_2CO_3	Carbonic acid

Naming Bases

Most bases are hydroxides: e.g. sodium hydroxide, potassium hydroxide etc.

Some other bases are carbonates and hydrogen carbonates (these are weaker bases) e.g.: sodium carbonate, sodium hydrogen carbonate etc.

Polyatomic Ions and Their Acids

Formula	Name	Oxoacid	Name
CO_3^{2-}	Carbonate	H_2CO_3	Carbonic acid
HCO_3^-	Hydrogen carbonate		
SO_4^{2-}	Sulfate	H_2SO_4	Sulfuric acid
HSO_4^-	Hydrogen sulfate		
SO_3^{2-}	Sulfite		
HSO_3^-	Hydrogen sulfite		
NO_3^-	Nitrate	HNO_3	Nitric acid
NO_2^-	Nitrite		
PO_4^{3-}	Phosphate	H_3PO_4	Phosphoric acid
HPO_4^{2-}	Hydrogen phosphate		
$H_2PO_4^-$	Dihydrogen phosphate		
$C_2H_3O_2^-$ CH_3COO-	Acetate	$HC_2H_3O_2$ CH_3COOH	Acetate

Naming Hydrates

- Hydrates are ionic compounds that are associated with water.
- The water gets trapped in the ions when the solid crystallizes, and in some cases the solids absorb water (hygroscopic)
- This water can be removed from the compound by heating the compound.
- Naming is as usual for the ionic compound and then add the water as hydrate with the appropriate prefix of the number of water molecules.

Examples and Names:
$FeSO_4 \cdot 7H_2O$ - Iron(II) sulfate heptahydrate
$CuSO_4 \cdot 5H_2O$ - Copper (II) sulfate pentahydrate

Some Compounds with Common Names

Formula	Common name	IUPAC Name
H_2O	Water	Dihydrogen monoxide
NH_3	Ammonia	Trihydrogen nitride
CO_2	Dry ice	Carbon dioxide (solid)
NaCl	Salt	Sodium chloride
$CaCO_3$	Marble, chalk	Calcium carbonate
$MgSO_4 \cdot 7H_2O$	Epsom salt	Magnesium sulfate heptahydrate
$Mg(OH)_2$	Milk of magnesia	Magnesium hydroxide
CH_4	Methane	methane

Notes:

Notes:

Chapter 2: Atoms, Molecules and Ions

Keywords/Concepts

- Empirical formula
- Molecular formula
- Structural formula
- Formula unit
- Ionic compounds nomenclature
- Covalent compounds nomenclature
- Transition metal ions
- Polyatomic ions
- Acids and bases
- Hydrates

Notes:

Ch 2/ PowerPoint Study–02-2Nomenclature   Name: _____

Answer these questions as you are watching the videos. They are due in class.
These questions are not just for you to answer but also to prepare you for the exam. So make sure you understand what you are writing and not just copy from the text book. **Show all work.**
For the following compounds fill the columns. I have solved a couple to give you an idea.

Name/formula	Covalent/Ionic compound?	Cation, if present. Write the transition metal or polyatomic ion, if present.	Anion, if present. Write the polyatomic ion, if present.	Name/Formula
$CaCl_2$				
N_2O_4				
$Ca(NO_3)_2$				
HF				
$Co(OH)_2$	*Ionic*	*Co^{2+}, transition metal*	*OH^-, polyatomic hydroxide*	*Cobalt (II) hydroxide*
CuCl				
NH_4NO_2				
$FeSO_4$				
potassium carbonate				
carbon disulfide	*Covalent*	*None*	*None*	*CS_2*
hydrogen chloride				
tin(II) nitride				
lead(IV) phosphate				
calcium hydride				

Ch 2/ Worksheet/Atomic Structure   Name: _____

1. Write three aspects of Dalton's atomic theory.

2. Fill in the following table: (*negative charge means electrons added; positive means electrons removed*)

Nuclear Symbol	Atomic Number	Mass Number	Number of Protons	Number of Electrons	Number of Neutrons	Charge Possible
$^{40}_{18}Ar$	_____	_____	_____	_____	_____	_____
_____	_____	39	19	18	_____	_____
_____	16	_____	_____	_____	20	-2

3. Provide the common names of Groups 1, 2, 7 and 8.

4. For each of the following elements, indicate whether it is a main group element (MG), or transition metal (TM). If the element is a main group element, indicate the group number and whether it is a metal, a nonmetal or a metalloid. Write the symbol and/or name of the element. Write the charge possible on the ion (don't write for TM).

Sr	Bromine	Mo
P	Magnesium	B
Lead	Hg	Mn

Ch 2/ Worksheet/Nomenclature-1   Name: _____

1. Write the ions that can form from the following elements and the names. An example is given in the first row.

Na⁺	Sodium ion		iodide	I⁻
Fe²⁺				P³⁻
	Calcium ion		Oxide	
Cr⁶⁺			Hydride	

2. Write the names or give the formulas of the following polyatomic ions.

OH⁻				Carbonate
	Phosphate		HCO₃⁻	
NO₃⁻				Hydroxide
	Ammonium ion		SO₄²⁻	

3. Classify the following as covalent or ionic compounds

P₂O₅	Iron (II) Oxide	Sodium oxide
Calcium chloride	NO₃	CO₂
BaO	AlCl₃	MnS₂

4. Give four diatomic compounds with the same name as their element name.

Chapter 2: Atoms, Molecules and Ions

5. Identify and name the polyatomic ions in the following compounds.

Na_2SO_4	KNO_3	NH_4Cl
$Ca_3(PO_4)_2$	NH_4NO_3	$Sr(OH)_2$

6. Give the names or formulas of the following compounds.

K_3N		silver bromide	
SO_2		silicon dioxide	
$Pb(SO_4)_2$		carbon tetrachloride	
$Fe(NO_3)_3$		lead (II) nitride	
$Al(CN)_3$		tin (II) nitrite	
$Mn_2(SO_3)_3$		cobalt (III) oxide	
$SnSe_2$		chromium (III) hydroxide	
$Be(HCO_3)_2$		titanium (II) acetate	
$CuOH$		magnesium sulfate heptahydrate	
NH_4Cl		potassium carbonate	
Cu_3P		diboron tetrabromide	
$Ca(C_2H_3O_2)_2$		lithium iodide	
$FePO_4$		silver acetate	
$NaBr$		manganese (II) phosphate	
P_2O_5		chromium (VI) phosphate	
$Zn(NO_2)_2$		vanadium (V) sulfide	
		nickel (III) sulfide	

Ch 2/ Worksheet/Nomenclature – Acids/ Bases/Hydrates

Name: _____

1. Name the following binary acids:
   a. HF
   b. HBr
   c. HI

2. Name the following oxoacids:
   a. H_2SO_4
   b. H_2CO_3
   c. $HClO_3$

   d. HClO
   e. HNO_3
   f. H_2SO_3

3. Name the following bases:
   a. $Mg(OH)_2$
   b. LiOH
   c. $Ba(OH)_2$

4. Give the formula for the following acids or bases:
   a. perchloric acid
   b. phosphoric acid
   c. nitrous acid

   d. hydrochloric acid
   e. nitric acid
   f. ammonium hydroxide

   g. copper (II) hydroxide
   h. zinc hydroxide
   i. rubidium hydroxide

5. Give the name or structure of the following hydrates:
   a. Magnesium sulfate heptahydrate
   b. $CuSO_4 \cdot 2H_2O$

   c. Cobalt (II) chloride hexahydrate

Ch 2/ Worksheet/Nomenclature-2   Name: _____

1. Ionic Compounds – Formula

Name	Cation	Anion	Formula
Sodium chloride			
Magnesium chloride			
Calcium oxide			
Lithium phosphide			
Aluminum sulfide			
Calcium nitride			

2. Ionic Compounds – Names

Formula	Cation	Anion	Name
K_2S			
BaF_2			
MgO			
Na_3N			
$AlCl_3$			
Mg_3P_2			

3. Ionic Compounds (Transition Metals) – Formula

Name	Cation	Anion	Formula
Iron (III) chloride			
Iron (II) oxide			
Copper (I) sulfide			
Copper (II) nitride			
Zinc oxide			
Silver sulfide			

4. Ionic Compounds (Transition Metals) – Name

Formula	Cation	Anion	Name
Cu_2S			
Fe_2O_3			
$CuCl_2$			
FeS			
Ag_2O			
$FeBr_2$			

5. Ionic Compounds (Polyatomic Ions) – Formula

Name	Cation	Anion	Formula
Potassium carbonate			
Sodium nitrate			
Calcium hydrogen carbonate			
Manganese (II) sulfate			
Chromium (III) nitrite			
Lithium phosphate			

6. Ionic Compounds (Polyatomic Ions) – Name

Formula	Cation	Anion	Name
$CaSO_4$			
$Al(NO_3)_3$			
Na_2CO_3			
$MgSO_3$			
$Cu(OH)_2$			
$Mg_3(PO_4)_2$			

7. Covalent Compounds – Formula

Name	Formula
Sulfur trioxide	
Dinitrogen monoxide	
Iodine pentachloride	
Diphosphorous pentaoxide	
Dihydrogen sulfide	
Sulfur diflouride	

8. Covalent Compounds – Names

Formula	Name
N_2O_5	
PCl_3	
$BrCl_3$	
CS_2	
BF_3	
XeI_2	

Chapter 3
Stoichiometry: Chemical Calculations

Atomic book keeping
Calculation Atomic mass and molecular/formula mass: separate all the atoms in the molecular formula and find atomic mass from periodic table and add the mass to obtain formula/molecular mass.

The Mole

1 mole = 6.02×10^{23} atoms = atomic mass of element (from the PT)
1 mol of Au atoms = 6.02×10^{23} atoms = 196.9665 g/mol
1 mol of Cl_2 = 6.02×10^{23} Cl_2 atoms = 79.906g /mol of Cl_2
1 mol of $AlCl_3$ = 6.02×10^{23} atoms
1 mol Al^{3+} ions and 3 mols Cl^- ions.

Mass percent composition

1. see molecular formula
2. add up all similar atoms
3. calculate mass of the different atom(s)
4. calculate formula mass
5. divide mass of different atoms by formula mass and multiply by 100%
6. add all %s to make sure you get hundred (there should be no other units left)

Other calculations using moles e.g. mass of an element in a given compound.
Elemental Analysis and Calculation of molecular formula.
Review: empirical formula and relationship to molecular formula.

Stoichiometry

Writing and balancing chemical equations
Reactant(s) $\longrightarrow$ product(s)

   Solid (s), liquid (l), gas (g), aqueous (aq)

Coefficient – the number in front of compound or element after balancing the equation.
Calculations using stoichiometry:

1. write equation
2. balance equation
3. write quantities given under compounds/elements
4. start with what you know!!!!
   a. Calculate mols of given quantity
   b. Find the mol ratio of given to needed from the balanced equation
   c. Convert mol to gram of the answer.
      e.g.
      A + 2B $\longrightarrow$ 2C + 3D
      2 g   ? g

      $$\frac{2 \text{ g } A}{} \times \frac{1 \text{ mol } A}{g \text{ } A} \times \frac{2 \text{ mol } C}{1 \text{ mol } A} \times \frac{g \text{ } C}{1 \text{ mol } C} = g \text{ } C$$

Limiting reagent: need to calculate mols of all starting materials to find out which is less, that will be the limiting reagent.
Percent yields:
   Percent Yield = actual/theoretical x 100%
(Theoretical – from stoichiometric calculations and Actual – after performing experiment in the lab)

Chapter 3 Mass Percent and Empirical Formula

Dr. Sapna Gupta

Molecular and Formula Mass

- Molecular mass is for molecules (covalent compounds)
- Formula mass is for ionic compounds.
- Masses are calculated by adding the mass of all the elements in the formula.
- It is really important to be good in nomenclature by now. The most common error is not knowing the formula of the compound.

Notes:

Solved Examples

Calculate the formula weight of the following compounds from their formulas. Report your answers to three significant figures.

Calcium hydroxide, $Ca(OH)_2$
1 Ca	1(40.08) =	40.08 amu
2 O	2(16.00) =	32.00 amu
2 H	2(1.008) =	2.016 amu
Total		74.096 amu; correct SF = 74.10 amu

Methylamine, CH_3NH_2
1 C	1(12.01) =	12.01 amu
1 N	1(14.01) =	14.01 amu
5 H	5(1.008) =	5.040 amu
Total		31.060 amu; correct SF = 31.06 amu

Nitric acid, HNO_3
1 H	1(1.008) =	1.008 amu
1 N	1(14.01) =	14.01 amu
3 O	3(16.00) =	48.00 amu
Total		63.018 amu; correct SF = 63.02 amu

Notes:

Mole Concept

Question: Which has more number of atoms? 1 g of He or 1 g of Au?

Answer: He – it will take many more atoms of helium to weigh 1 g because the mass of He atom is about 4 amu (or grams for convenience) per atom, whereas Au is 197g per atom.

In chemical reactions atoms undergo rearrangement to form new compounds. There is also no loss of matter (Law of Conservation of Matter). So we have to account for ALL atoms. To calculate how many atoms will be completely going from one rearrangement to another, we need to be able to count them.

Avogadro, an Italian chemist, was the first one to calculate a number using mass of protons and carbon -12 isotope that could be useful for such calculations.

Mole Concept (mol)

- Mole is a measurement given to a certain number of atoms.
- 1 mol of substance has 6.022×10^{23} number of atoms
- 1 mol is like 1 dozen – a conversion factor:
    - 1 mol of people are 6.022×10^{23} of people
    - 1 mol of pencils are 6.022×10^{23} number of pencils
    - 1 mol of stars are 6.022×10^{23} number of stars
- The mass of 1 mol of a substance is the mass number of the substance.
- Example:
    1 mol C = 12.01 g of C = 6.022×10^{23} atoms of C
    1 mol Fe = 55.847 g Fe = 6.022×10^{23} atoms of Fe

Atomic Mass

- Units of mass on the periodic table are reported as g/mol.
- Nitrogen is 14.007 g/mol – there are 14.007 g of N in every mol of N
- Sodium is 22.989 g/mol – there are 22.989 g of Na in every mol of Na
- For majority of calculations we don't use the number of atoms as they cannot be counted. It is easier to work in grams.
- Remember – mols is a unit but it cannot be measured by any instrument; it has to be converted to grams to be measured.
- The number of atoms are important because all reactions take place at atomic level and according to the conservation of mass no atoms are destroyed – only rearranged. Mol is an easy way to work with atoms without working in exponents.

Notes:

Molar Mass

- Atomic mass is obtained from the periodic table – the mass numbers given are what is used.
- Molar mass (and formula mass) is the addition of the mass of ALL the atoms in the formula.
- Example: as before calculate the mass of calcium hydroxide. Use the same numbers and method, but instead of using amu – use g/mol.

Calcium hydroxide, $Ca(OH)_2$

1 Ca	1(40.08) =	40.08 g/mol
2 O	2(16.00) =	32.00 g/mol
2 H	2(1.008) =	2.016 g/mol
Total		74.096 g/mol; correct SF = 74.10 g/mol

Notes:

Chapter 3: Stoichiometry: Chemical Calculations

Solved Examples

- You have to be able to covert number of atoms to mols, mols to grams and all possible combinations.
- These calculations are done similar to unit conversion as in chapter 1; i.e. dimensional analysis.

Example:

A sample of nitric acid, HNO_3, contains 0.253 mol HNO_3. How many grams is this?

First, find the molar mass of HNO_3:

1 H	1(1.008) =	1.008
1 N	1(14.01) =	14.01
3 O	3(16.00) =	48.00
Total		63.018 g/mol

$$\frac{0.253 \, mol}{} \times \frac{63.02 \, g}{1 \, mol} = 15.94406 \, g = 15.9 \, g/mol$$

Calcite is a mineral composed of calcium carbonate, $CaCO_3$. A sample of calcite composed of pure calcium carbonate weighs 23.6 g. How many moles of calcium carbonate is this?

First, find the molar mass of $CaCO_3$:

1 Ca	1(40.08) =	40.08	
1 C	1(12.01) =	12.01	
3 O	3(16.00) =	48.00	2 decimal places
		100.09	100.09 g/mol

Next, find the number of moles in 23.6 g:

$$23.6 \, g \times \frac{1 \, mole}{100.09 \, g} = 2.35787791 \times 10^{-1} \, g$$

$$= 2.36 \times 10^{-1} \, g \text{ or } 0.236 \, g$$
(3 significant figures)

Notes:

Chapter 3: Stoichiometry: Chemical Calculations

The average daily requirement of the essential amino acid leucine, $C_6H_{14}O_2N$, is 2.2 g for an adult. What is the average daily requirement of leucine in moles?

First, find the molar mass of leucine:

6 C	6(12.01) =	72.06
2 O	2(16.00) =	32.00
1 N	1(14.01) =	14.01
14 H	14(1.008) =	14.112
		132.182

2 decimal places
132.18 g/mol

Next, find the number of moles in 2.2 g:

$$2.2 \text{ g} \times \frac{1 \text{ mole}}{132.18 \text{ g}} = 1.6643 \times 10^{-2} \text{ mol} = 1.7 \times 10^{-2} \text{ mol} \text{ or } 0.017 \text{ mol}$$
(2 significant figures)

Notes:

The daily requirement of chromium in the human diet is 1.0×10^{-6} g. How many atoms of chromium does this represent?

First, find the molar mass of Cr:
1 Cr   1(51.996) = 51.996

Now, convert 1.0×10^{-6} grams to moles:

$$1.0 \times 10^{-6} \text{ g} \times \frac{1 \text{ mol}}{51.996 \text{ g}} \times \frac{6.022 \times 10^{23} \text{ atoms}}{1 \text{ mol}} = 1.158166 \times 10^{16} \text{ atoms}$$

1.2×10^{16} atoms
(2 significant figures)

Notes:

Chapter 3: Stoichiometry: Chemical Calculations

Mass Percents

- Mass percent of compounds is the mass percent of each element in that compound.
- Find the total mass of the compound and find the mass of each of the elements in the compound.

$$\% \text{ mass of an element} = \frac{\text{mols of element} \times \text{at.mass of element}}{\text{molar mass of the compound}} \times 100\%$$

e.g. in Na_2CO_3 – there are 2 mols of sodium, 1 mol of carbon and 3 mols of oxygen.

Notes:

Lead(II) chromate, $PbCrO_4$, is used as a paint pigment (chrome yellow). What is the percentage composition of lead(II) chromate?

First, find the molar mass of $PbCrO_4$:

1 Pb	1(207.2) =	207.2
1 Cr	1(51.996) =	51.996
4 O	4(16.00) =	64.00

323.196   (1 decimal place)
323.2 g/mol

Now, convert each to percent composition:

Pb: $\dfrac{207.2 \text{ g}}{323.20 \text{ g}} \times 100\% = 64.11\%$

Cr: $\dfrac{51.996 \text{ g}}{323.20 \text{ g}} \times 100\% = 16.09\%$

Check:
$64.11 + 16.09 + 19.80 = 100.00$

O: $\dfrac{64.00 \text{ g}}{323.20 \text{ g}} \times 100\% = 19.80\%$

Notes:

Chapter 3: Stoichiometry: Chemical Calculations

The chemical name of table sugar is sucrose, $C_{12}H_{22}O_{11}$. How many grams of carbon are in 68.1 g of sucrose?

First, find the molar mass of $C_{12}H_{22}O_{11}$:

12 C	12(12.01) =	144.12
11 O	11(16.00) =	176.00
22 H	22(1.008) =	22.176 (2 decimal places)
		342.296 342.30 g/mol

Now, find the mass of carbon in 68.1 g sucrose:

$$68.1 \text{ g sucrose} \times \frac{144.12 \text{ g carbon}}{342.30 \text{ g sucrose}} = 28.7 \text{ g carbon}$$

(3 significant figures)

Notes:

What is the percent water in copper(II) sulfate pentahydrate, $CuSO_4 \cdot 5 H_2O$?

1 Cu	(1)63.55 g = 63.55 g	2 H	(2)1.01 = 2.02 g
1 S	(1)32.07 g = 32.07 g	1 O	(1)16.00 = 16.00 g
4 O	(4)16.00 g = 64.00 g		18.02 g/mol
	159.62 g/mol		

Formula Mass = 159.62 + 5(18.02) = 249.72 g/mol

Divide the mass of water in one mole of the hydrate by the molar mass of the hydrate and multiply this fraction by 100.

$$\text{percent hydration} = \frac{(5 \times 18.02)}{249.72} \times 100\% = 36.08\%$$

Notes:

More Examples: Converting Mass, Moles and Atoms

Example 1: Determine the number of moles in 85.00 grams of sodium chlorate, $NaClO_3$

$$85.00 \text{ g } NaClO_3 \times \frac{1 \text{ mole } NaClO_3}{106.44 \text{ g } NaClO_3} = 0.7986 \text{ mol } NaClO_3$$

Example 2: Determine the number of molecules in 4.6 moles of ethanol, C_2H_5OH. (1 mole = 6.022×10^{23})

$$4.6 \text{ mol } C_2H_5OH \times \frac{6.02 \times 10^{23} \text{ molecules } C_2H_5OH}{1 \text{ mol } C_2H_5OH} = 2.8 \times 10^{24} \text{ molecules}$$

Example 3: (continued from 2) Determine how many H atoms are in 4.6 moles of ethanol.
- Begin with the answer to the last problem

$$2.8 \times 10^{24} \text{ molecules } C_2H_5OH \times \frac{6 \text{ H atoms}}{1 \text{ molecule } C_2H_5OH} = 1.7 \times 10^{25} \text{ atoms H}$$

Notes:

Combustion Analysis

- Analysis of organic compounds (C, H, N, S and sometimes O) are carried using an apparatus like the one below.

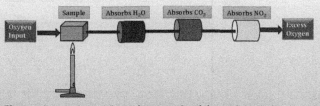

- The sample is combusted (burned in oxygen) and the products, carbon dioxide and water and other oxides are collected and weighed.
- The percent of elements can then be calculated followed by calculation of the empirical formula.

Source: Sapna Gupta

Notes:

Empirical Formula

The formula of a substance written with the smallest integer subscripts.
  The empirical formula for N_2O_4 is NO_2.
  The empirical formula for H_2O_2 is HO.

Determining the Empirical Formula

Beginning with percent composition:
- Assume exactly 100 g so percentages convert directly to grams.
- Convert grams to moles for each element.
- Round off the resulting mole ratios to obtain whole numbers.
- Divide each mole amount by the smallest mole amount to get mol ratio.

If the result is not a whole number:
- Multiply each mole amount by a factor to make whole numbers.

For example:
- If the decimal portion is 0.5, multiply by 2.
- If the decimal portion is 0.33 or 0.67, multiply by 3.
- If the decimal portion is 0.25 or 0.75, multiply by 4.

To get the molecular formula divide the molecular weight by the mass of the empirical formula. Multiply the empirical formula with that number obtained.

Benzene is composed of 92.3% carbon and 7.7% hydrogen. What is the empirical formula of benzene? Its molecular weight is 78.1 amu. What is its molecular formula?

$$92.3 \text{ g C} \cdot \frac{1 \text{ mol C}}{12.01 \text{ g C}} = 7.685 \text{ mol C} \qquad \frac{7.685}{7.64} = 1$$

$$7.7 \text{ g H} \cdot \frac{1 \text{ mol H}}{1.008 \text{ g H}} = 7.64 \text{ mol H} \qquad \frac{7.64}{7.64} = 1$$

Empirical formula: CH

Empirical formula weight = 13.02 amu

$$\frac{78.1}{13.02} = 6$$

Molecular formula: C_6H_6

Chapter 3: Stoichiometry: Chemical Calculations

Sodium pyrophosphate is used in detergent preparations. It is composed of 34.5% Na, 23.3% P, and 42.1% O. What is its empirical formula?

Step 1

$$34.5 \text{ g Na} \times \frac{1 \text{ mol Na}}{22.99 \text{ g Na}} = 1.501 \text{ mol Na}$$

$$23.3 \text{ g P} \times \frac{1 \text{ mol P}}{30.97 \text{ g P}} = 0.7523 \text{ mol P}$$

$$42.1 \text{ g O} \times \frac{1 \text{ mol O}}{16.00 \text{ g O}} = 2.631 \text{ mol O}$$

Step 2

1.501 mol Na $\frac{1.501}{0.7523} = 2.00 \quad \times 2 = 4$

0.7523 mol P $\frac{0.7523}{0.7523} = 1.00 \quad \times 2 = 2$

2.631 mol O $\frac{2.631}{0.7523} = 3.50 \quad \times 2 = 7$

Empirical formula $Na_4P_2O_7$

Notes:

Hexamethylene is one of the materials used to produce a type of nylon. It is composed of 62.1% C, 13.8% H, and 24.1% N. Its molecular weight is 116 amu. What is its molecular formula?

Step 1

$$62.1 \text{ g C} \times \frac{1 \text{ mol C}}{12.01 \text{ g C}} = 5.171 \text{ mol C}$$

$$13.8 \text{ g H} \times \frac{1 \text{ mol H}}{1.008 \text{ g H}} = 13.69 \text{ mol H}$$

$$24.1 \text{ g N} \times \frac{1 \text{ mol N}}{14.01 \text{ g N}} = 1.720 \text{ mol H}$$

Step 2

5.171 mol C $\frac{5.171}{1.720} = 3$

13.69 mol H $\frac{13.69}{1.720} = 8$

1.720 mol H $\frac{1.720}{1.720} = 1$

Step 3

Empirical formula C_3H_8N

Step 4

Find the formula weight of the empirical formula C_3H_8N.

$3(12.01) + 8(1.008) + 1(14.01) = 58.104$ amu

$$n = \frac{116}{58.10} = 2$$

Molecular formula: $C_6H_{16}N_2$

Notes:

Chapter 3: Stoichiometry: Chemical Calculations

Key Words and Concepts

- Formula mass
- Mole concept
- Avogadro's number
- Mass percents
- Elemental analysis

Notes:

Ch 3/ PowerPoint Study-1 MolarMass/Moles/Atoms   Name: _____

Answer these questions as you are watching the videos. They are due in class.
These questions are not just for you to answer but also to prepare you for the exam.
Make sure you understand what you are writing and not just copy from the text book. **Show all work.**

1. Calculate the molar mass of the following compounds:
   a. Na_2CO_3 *(2 Na+1 C+3 O)*

   b. Co_3N_2

   c. copper (I) arsenide

   d. diphospohorous pentaoxide

2. How many mols are in the following (you can use the molar masses from question 1):
   a. 1.33 g of Co_3N_2

   b. 2.45 g of copper (I) arsenide

3. How many grams are in the following:
   a. 6.55 mols of Na_2CO_3

   b. 6.55×10^{22} molecules of potassium oxide

Chapter 3: Stoichiometry: Chemical Calculations

4. How many molecules are in the following:
   a. 10.2 moles of diphosphorous pentaoxide

   b. 9.66 grams of potassium oxide

5. How many oxygen atoms are in 3.55 g of Na_2CO_3.

6. What is mass percent of each element in copper (I) carbonate?

7. What is the mass percent of water in magnesium sulfate heptahydrate?

8. What is the empirical formula of a compound containing 77.7% iron and 22.3 % oxygen?

Chapter 3 Stoichiometry - Equations   *Dr. Sapna Gupta*

Stoichiometry

- **Stoichiometry** helps us to find out
  - How much starting material is required to produce a certain amount of product
  - The amount of product that can be produced from a certain amount of starting material.
  - How will the reaction be affected if there is more than one starting material (limiting reagent)?
  - Will there be any starting material left over?
  - How efficient is the process (% yield)?
- A **chemical reaction** is representation of chemicals in a reaction.
  - A reaction is written in chemical symbols so that it is clear how many atoms are being used.
  - A chemical reaction where reactants are written on the left and products on the right with an arrow (yield) to show progress of reaction.

$$\text{Reactants} \xrightarrow{\text{yields}} \text{Products}$$

Notes:

Chemical Equations

$$\text{Reactants} \longrightarrow \text{Products}$$

Here is an example:
$$C + O_2 \longrightarrow CO_2$$

- The law of conservation of mass has to be obeyed.
- The number of atoms on the reactant side should be equal to the atoms on the right side.
- Number of C atoms is one on both sides and number of oxygen atoms is two.
- They are not in their original form (chemical change) but all atoms are accounted for.
- This equation is balanced. If atoms are not accounted for then we have to balance the chemical equation.

Notes:

Chapter 3: Stoichiometry: Chemical Calculations

Balancing Chemical Equations

For example, the reaction of sodium with chlorine produced sodium chloride.

- First, we determine the correct formula for each compound.
  - Sodium is Na; Chlorine is Cl_2; Sodium chloride is NaCl.
- Second, we write the reaction: $Na + Cl_2 \rightarrow NaCl$
- Third, we check the number of each atom on each side of the equation.
- The equation shows two Cl atoms on the reactant side and only one Cl atom on the product side. To balance the Cl atoms, we insert a coefficient of "2" before NaCl on the product side.

$$Na + Cl_2 \rightarrow 2NaCl$$

- Now the Na are not balanced: there is one on the reactant side and there are two on the product side. To balance Na, we insert the coefficient "2" before Na on the reactant side.

$$2Na + Cl_2 \rightarrow 2NaCl$$

Notes:

Word Equation to Symbol Equation

- It is common to write equations in word format that have to be then converted into a proper chemical equation using chemical symbols, like shown in the previous slide.
- Hints to writing equations:
  - Metals are always monoatomic, e.g. potassium (K), tin (Sn)
  - Some elements exist only as diatomic gases, e.g. chlorine (Cl_2), oxygen (O_2), hydrogen (H_2). In an chemical reaction, this has to be remembered.
  - All compounds should be written with proper mol ratios e.g. sodium chloride (NaCl), calcium chloride ($CaCl_2$), magnesium oxide (MgO). Be good in nomenclature.
  - Read from the problem what are the products and what are the reactants.

Example: Write the equation for magnesium reacting with nitrogen to give magnesium nitride.
1) Symbols for all chemicals: Mg, N_2, Mg_3N_2.
2) Write the equation: what are the reactants? Mg and N_2
   Magnesium + nitrogen → magnesium nitride
   $Mg + N_2 \rightarrow Mg_3N_2$
3) Now balance the equation: leave Mg for last because it is by itself.
   $Mg + 2N_2 \rightarrow 2Mg_3N_2$
   Finally balance Mg:
   $6Mg + 2N_2 \rightarrow 2Mg_3N_2$

Notes:

Chapter 3: Stoichiometry: Chemical Calculations   85

Balance the following equation: $CS_2 + O_2 \rightarrow CO_2 + SO_2$

Tally the number of each atom on each side:
- C — 1 on reactant side; 1 on product side
- S — 2 on reactant side; 1 on product side
- O — 2 on reactant side; 4 on product side

Begin by inserting the coefficient "2" before SO_2 on the product side. We leave O_2 until later because it is an element.

$$CS_2 + O_2 \rightarrow CO_2 + 2SO_2$$

Tally the atoms again:
- C — 1 on reactant side; 1 on product side
- S — 2 on reactant side; 2 on product side
- O — 2 on reactant side; 6 on product side

Insert a "3" before O_2:

$$CS_2 + 3O_2 \rightarrow CO_2 + 2SO_2$$

Tally the atoms again:
- C — 1 on reactant side; 1 on product side
- S — 2 on reactant side; 2 on product side
- O — 6 on reactant side; 6 on product side

Notes:

Balance the following equation: $NH_3 + O_2 \rightarrow NO + H_2O$

1) Tally the number of each atom on each side:
- N — 1 on reactant side; 1 on product side
- H — 3 on reactant side; 2 on product side
- O — 2 on reactant side; 2 on product side

2) Begin by inserting the coefficient "2" before NH_3 on the reactant side and the coefficient "3" before H_2O on the product side. We leave O_2 until later because it is an element.

$$2NH_3 + O_2 \rightarrow NO + 3H_2O$$

3) Tally the atoms again:
- N — 2 on reactant side; 1 on product side
- H — 6 on reactant side; 6 on product side
- O — 2 on reactant side; 4 on product side

4) To balance N, insert a "2" before NO:

$$2NH_3 + O_2 \rightarrow 2NO + 3H_2O$$

5) Tally the atoms again:
- N — 2 on reactant side; 2 on product side
- H — 6 on reactant side; 6 on product side
- O — 2 on reactant side; 5 on product side

6) Since Since this gives us an odd number oxygens, we double the coefficients on NH_3, NO, and H_2O and to balance O, insert a "5" before O_2.

7) Tally the atoms again to double check:

$$4NH_3 + 5O_2 \rightarrow 4NO + 6H_2O$$

- N — 4 on reactant side; 4 on product side
- H — 12 on reactant side; 12 on product side
- O — 10 on reactant side; 10 on product side

The reaction is now balanced!

Notes:

Types of Reactions

- **Synthesis** (combination): two substances combine to form one.
  $$2Na(s) + Cl_2(g) \rightarrow 2NaCl(s)$$
- **Double Displacement**: A reaction in which two elements displaces two elements.
  $$AgNO_3(aq) + NaCl(aq) \rightarrow AgCl(s) + NaNO_3(aq)$$
- **Single displacement**: A reaction where one element displaces one other element.
  $$Zn(s) + CuSO_4(aq) \rightarrow ZnSO_4(aq) + Cu(s)$$
- **Decomposition**: A reaction in which a single compound reacts to give two or more substances.
  $$2HgO(s) \rightarrow 2Hg(l) + O_2(g)$$

(g) – gas; (l) – liquid; (s) – solid; (aq) – dissolved in water

Notes:

Review

- Chemical equations
  - Reactants
  - Products
  - State symbols
  - Balancing
- Writing chemical equation from word equations
- Types of reactions
  - Synthesis
  - Double displacement
  - Single displacement
  - Decomposition

Notes:

Ch 3/ PowerPoint Study-2 Stoichiometry – Equations   Name: _____

Answer these questions as you are watching the videos. They are due in class.
These questions are not just for you to answer but also to prepare you for the exam.
Make sure you understand what you are writing and not just copy from the text book. **Show all work.**

1. Balance the following equations:

    a. $NH_4NO_3 \rightarrow N_2O + H_2O$

    b. $Mg + CO_2 \rightarrow MgO + C$

    c. $NaCl + Pb(NO_3)_2 \rightarrow PbCl_2 + NaNO_3$

2. Classify the following equations as synthesis, double displacement, single displacement, or decomposition.

    a. $(NH_4)_2Cr_2O_7 \rightarrow N_2 + 4H_2O + Cr_2O_3$ _____

    b. $12C + 11H_2O \rightarrow C_{12}H_{22}O_{11}$ _____

    c. $2HCl + Ba(OH)_2 \rightarrow BaCl_2 + 2H_2O$ _____

An equation can be written from a word equation. For example:

Word equation: zinc sulfide reacts with oxygen to form zinc oxide and sulfur

Chemical equation: $ZnS + O_2 \longrightarrow ZnO + S$
Note:

   i. oxygen is diatomic element hence written as O_2.

   ii. Zn is a transition metal but will always has a valency of +2 hence (II) was not written in the word equation.

   iii. sulfur is not diatomic element, hence written as S.

Balanced equation: $2ZnS + O_2 \longrightarrow 2ZnO + 2S$

Write the following word equations in chemical equation, and then balance the equations.

Word equation: barium hydroxide and sulfuric acid make water and barium sulfate

Balanced Chemical equation

Word equation: copper metal and silver nitrate react to form silver metal and copper (II) nitrate

Balanced Chemical equation

Chapter 3   Stoichiometry - Introduction

Dr. Sapna Gupta

Stoichiometry

- **Stoichiometry** helps us to find out
  - How much starting material is required to produce a certain amount of product
  - The amount of product that can be produced from a certain amount of starting material.
  - How will the reaction be affected if there is more than one starting material (limiting reagent)?
  - Will there be any starting material left over?
  - How efficient is the process (% yield)?
- A **chemical reaction** is representation of chemicals in a reaction.
  - A reaction is written in chemical symbols so that it is clear how many atoms are being used.
  - A chemical reaction where reactants are written on the left and products on the right with an arrow (yield) to show progress of reaction.

Reactants $\xrightarrow{\text{yields}}$ Products

Notes:

Stoichiometry – setting up

The calculation of the quantities of reactants and products involved in a chemical reaction.

Interpreting a Chemical Equation

The coefficients of the balanced chemical equation can be interpreted as either
(1) numbers of molecules (or ions or formula units) or
(2) numbers of moles, depending on your needs.

$$2\ H_{2(g)} + O_{2(g)} \rightarrow 2\ H_2O_{(l)}$$

2 molecules $H_{2(g)}$ + 1 molecule $O_{2(g)}$ → 2 molecules $H_2O_{(l)}$

2 moles $H_{2(g)}$ + 1 mole $O_{2(g)}$ → 2 moles $H_2O_{(l)}$

This relationship is made because of Avogadro's number (N_A)

Notes:

Chapter 3: Stoichiometry: Chemical Calculations

Stoichiometry – setting up....

A ⟶ B

To find the amount of B (one reactant or product) given the amount of A (another reactant or product):

1. Convert grams of A to moles of A → Using the molar mass of A
2. Convert moles of A to moles of B → Using the coefficients of the balanced chemical equation
3. Convert moles of B to grams of B → Using the molar mass of B

Grams of A	Mols of A	Mols of B	Grams of B
	Find atomic/molar mass of A from periodic table	Find mol ratio of A to B from balanced chemical equation	Find atomic/molar mass of B from periodic table

Notes:

Solved Problem:
Propane, C_3H_8, is normally a gas, but it is sold as a fuel compressed as a liquid in steel cylinders. The gas burns according to the following equation:

$$C_3H_8(g) + 5O_2(g) \rightarrow 3CO_2(g) + 4H_2O(g)$$

How many grams of CO_2 are produced when 2.00 mols of propane are burned?

1) The equation is balanced.
2) Molar masses:
   C_3H_8: don't need mass for propane because mols are already given.
   CO_2: 1(12.01) + 2(16.00) = 44.01 g
3) Mol ratio needed is of C_3H_8 to CO_2, 1:3.
4) 1 mol of C_3H_8 produces 3 mols of CO_2.
5) Final step will be to covert mols CO_2 of to grams of CO_2.

$$2 \text{ mol } C_3H_8 \cdot \frac{3 \text{ mol } CO_2}{1 \text{ mol } C_3H_8} \cdot \frac{44.01 \text{ g } CO_2}{1 \text{ mol } CO_2} = 264 \text{ g } CO_2$$

Notes:

Chapter 3: Stoichiometry: Chemical Calculations

Solved Problem:

Propane, C_3H_8, is normally a gas, but it is sold as a fuel compressed as a liquid in steel cylinders. The gas burns according to the following equation:

$$C_3H_8(g) + 5O_2(g) \rightarrow 3CO_2(g) + 4H_2O(g)$$

How many grams of O_2 are required to burn 20.0 g of propane?

Molar masses:
O_2 $2(16.00) = 32.00$ g
C_3H_8 $3(12.01) + 8(1.008) = 44.094$ g

1) Convert g of propane to mol of propane using molar mass of propane.
2) Find the mol ratio of propane to oxygen to find mols of oxygen.
3) Find the grams of oxygen using molar mass of oxygen.

$$20.0 \text{ g } C_3H_8 \times \frac{1 \text{ mol } C_3H_8}{44.094 \text{ g } C_3H_8} \times \frac{5 \text{ mol } O_2}{1 \text{ mol } C_3H_8} \times \frac{32.00 \text{ g } O_2}{1 \text{ mol } O_2} = 72.57223205 \text{ g } O_2$$

$$\boxed{72.6 \text{ g } O_2}$$

Notes:

Stoichiometric Calculations - Mass to Mass

A chemist needs 58.75 grams of urea, how many grams of ammonia are needed to produce this amount?

$$2NH_3(g) + CO_2(g) \rightarrow (NH_2)_2CO(aq) + H_2O(l)$$

Strategy:
Grams urea → moles → mole ratio → grams

$$58.75 \text{ g } (NH_2)_2CO \times \frac{1 \text{ mol}(NH_2)_2CO}{58.06 \text{ g }(NH_2)_2CO} \times \frac{2 \text{ mol } NH_3}{1 \text{ mol }(NH_2)_2CO} \times \frac{17.04 \text{ g } NH_3}{1 \text{ mol } NH_3} = 34.49 \text{ g } NH_3$$

Notes:

Chapter 3: Stoichiometry: Chemical Calculations

Review

- Chemical equations
- Mole concept and conversions
- Stoichiometry

Notes:

Ch 3/ PowerPoint Study-3 Stoichiometry – Introduction

Name: _____

Answer these questions as you are watching the videos. They are due in class.
These questions are not just for you to answer but also to prepare you for the exam.
Make sure you understand what you are writing and not just copy from the text book. **Show all work.**

How many grams of sulfur can be produced from 2.00 mol of sulfur oxide? The equation is given below:
$$2H_2S + SO_2 \rightarrow 3S + 2H_2O$$

a. Is the equation balanced? Yes/No

b. Do you need the grams of sulfur dioxide? Yes/No

c. What is the mol ratio of sulfur dioxide to sulfur? _____

d. How many mols of sulfur will be formed from 2.00 mol of sulfur dioxide? _____

e. How many grams of sulfur are formed? Either convert your answer from (d) to grams of sulfur OR Do the dimensional analysis format for full stoichiometric setup.

Complete the following sentence with the information above.

2 mols of sulfur dioxide, will produce _____ mol of sulfur, which is _____ g of sulfur (theoretical yield).

A Second Scenario: Using the equation given above, try another problem.
How many grams of water will be formed when 3.00 g of sulfur are formed?

f. How many mols are in 3.00 g of sulfur?

g. What is the mol ratio of sulfur to water? _____

h. How many grams of water are produced? Again, either use the number from (g) to convert to grams or do the dimensional analysis format for full stoichiometric setup.

Chapter 3: Stoichiometry: Chemical Calculations

Stoichiometric Calculation – This is what a question looks like in the exam.
Do these problems using the dimensional analysis full setup instead of step by step.

Lithium is reacted with fluorine to form lithium fluoride as in the equation below.

$$Li + F_2 \longrightarrow LiF$$

1. How much lithium fluoride is made when 15.0 g of fluorine is used?

    (Strategy: a) check if the equation is balanced. If not, then balance it.

    b) convert 15 g F_2 → mol F_2 → mol ratio of LiF → g of LiF)

2. How much lithium is required to react with 23.2 g of fluorine? (follow the same strategy as above)

Chapter 3 Stoichiometry – 2 Limiting, Excess Reagent and Percent Yields *Dr. Sapna Gupta*

Limiting Reagent

- Any problem giving the starting amount for more than one reactant is a limiting reactant problem.
- The limiting reagent is entirely consumed when a reaction goes to completion.
- Once one reactant has been completely consumed, the reaction stops.
- **All amounts produced and reacted are determined by the limiting reactant.**
- How can we determine the limiting reactant?
  1. Use each given amount to calculate the amount of product produced.
  2. The limiting reactant will produce the *lesser* or *least* amount of product.

Magnesium metal is used to prepare zirconium metal, which is used to make the container for nuclear fuel (the nuclear fuel rods):

$$ZrCl_4(g) + 2Mg(s) \rightarrow 2MgCl_2(s) + Zr(s)$$

How many moles of zirconium metal can be produced from a reaction mixture containing 0.20 mol $ZrCl_4$ and 0.50 mol Mg?

$$0.20 \text{ mol } ZrCl_4 \cdot \frac{1 \text{ mol } Zr}{1 \text{ mol } ZrCl_4} = 0.20 \text{ mol } Zr$$

$$0.50 \text{ mol } Mg \cdot \frac{1 \text{ mol } Zr}{2 \text{ mol } Mg} = 0.25 \text{ mol } Zr$$

Since $ZrCl_4$ gives the lesser amount of Zr, $ZrCl_4$ is the limiting reactant.

0.20 mol Zr will be produced.

Chapter 3: Stoichiometry: Chemical Calculations

Urea, CH_4N_2O, is used as a nitrogen fertilizer. It is manufactured from ammonia and carbon dioxide at high pressure and high temperature:

$$2NH_3 + CO_2(g) \rightarrow CH_4N_2O + H_2O$$

In a laboratory experiment, 10.0 g NH_3 and 10.0 g CO_2 were added to a reaction vessel. What is the maximum quantity (in grams) of urea that can be obtained? How many grams of the excess reactant are left at the end of the reactions?

Molar masses
NH_3 1(14.01) + 3(1.008) = 17.02 g
CO_2 1(12.01) + 2(16.00) = 44.01 g
CH_4N_2O 1(12.01) + 4(1.008) + 2(14.01) + 1(16.00) = 60.06 g

$$10.0 \text{ g } NH_3 \times \frac{1 \text{ mol } NH_3}{17.024 \text{ g } NH_3} \times \frac{1 \text{ mol } CH_4N_2O}{2 \text{ mol } NH_3} \times \frac{60.06 \text{ g } CH_4N_2O}{1 \text{ mol } CH_4N_2O}$$
$$= 17.6 \text{ g } CH_4N_2O$$

$$10.0 \text{ g } CO_2 \times \frac{1 \text{ mol } CO_2}{44.01 \text{ g } CO_2} \times \frac{1 \text{ mol } CH_4N_2O}{1 \text{ mol } CO_2} \times \frac{60.06 \text{ g } CH_4N_2O}{1 \text{ mol } CH_4N_2O}$$
$$= 13.6 \text{ g } CH_4N_2O$$

CO_2 is the limiting reactant since it gives the lesser amount of CH_4N_2O. 13.6 g CH_4N_2O will be produced.

Notes:

To find the excess NH_3, we need to find how much NH_3 reacted. We use the limiting reactant as our starting point.

$$10.0 \text{ g } CO_2 \times \frac{1 \text{ mol } CO_2}{44.01 \text{ g } CO_2} \times \frac{2 \text{ mol } NH_3}{1 \text{ mol } CO_2} \times \frac{17.02 \text{ g } NH_3}{1 \text{ mol } NH_3}$$

$$= 7.73460577 \text{ g } NH_3$$

$$= 7.73 \text{ g } NH_3 \text{ reacted}$$

Now subtract the amount reacted from the starting amount:

```
  10.0     at start
 -7.73     reacted
  2.27 g   remains
```

2.27 g NH_3 is left unreacted

Notes:

Percent Yield

Theoretical Yield
Theoretical yield is the maximum amount of product that can be obtained by a reaction from given amounts of reactants.

This is a *calculated* amount.

Actual Yield
The amount of product that is actually obtained.

This is a *measured* amount.

Percentage Yield

$$\text{percentage yield} = \frac{\text{actual yield}}{\text{theoretical yield}} \times 100\%$$

Notes:

$$2NH_3 + CO_2(g) \rightarrow CH_4N_2O + H_2O$$

When 10.0 g NH_3 and 10.0 g CO_2 are added to a reaction vessel, the limiting reactant is CO_2. The theoretical yield is 13.6 of urea. When this reaction was carried out, 9.3 g of urea was obtained. What is the percent yield?

Theoretical yield = 13.6 g
Actual yield = 9.3 g

$$\text{Percent yield} = \frac{9.3 \text{ g}}{13.6 \text{ g}} \times 100\% \quad \boxed{= 68\% \text{ yield}}$$

Notes:

Chapter 3: Stoichiometry: Chemical Calculations

Review

- Limiting reactant
- Excess Reagent
- % yield

Notes:

Ch 3/ PowerPoint Study-3 Stoichiometry – Limiting Reagent/Excess

Name: _____

Answer these questions as you are watching the videos. They are due in class.
These questions are not just for you to answer but also to prepare you for the exam.
Make sure you understand what you are writing and not just copy from the text book. **Show all work.**

Calculating limiting reagent, percent yield and excess reagent.

The first step in the Ostwald process for producing nitric acid is
$$4NH_3(g) + 5O_2(g) \rightarrow 4NO(g) + 6H_2O(g)$$

a. If the reaction of 150. g of ammonia with 150. g of oxygen gas yields 87. g of nitric oxide (NO) then what is the percent yield of this reaction?
b. How much in grams is the excess reagent left?

Strategy: For solving part a

1. g NH_3 → mol NH_3 → mol ratio to NO

2. g O_2 → mol O_2 → mol ratio to NO

3. See which mols are less in steps 1 and 2. Less quantity means that is the starting material that will finish first and the mols calculated are the mols of the product formed.

4. Convert the lesser of the mols of NO to g of NO. This is the theoretical yield of NO.

5. Calculate % yield (87g NO divided by the grams (theoretical you calculated from step 4) and multiply by 100%.

Strategy: For solving part b

6. Subtract the smaller number from the larger number of mols from steps 1 and 2. This is mols of the reagent that is excess in quantity.

7. Mols of the excess product → mol ratio to the excess starting material → g of the starting material.

Ch 3/Worksheet/Naming, Molar Mass, Moles and Atoms Calculations

Name: _____

Show all work

Name (Write the name)	Formula (write the formula)	Molecular Mass	How many g, mols or molecules in
	$Be(C_2H_3O_2)_2$		g in 10.2 mols?
	NH_4OH		molecules in 9.66 g?
	N_2O_4		mols in 4.50×10^{25} molecules
ammonium sulfate			g in 6.55 mols?

Chapter 3: Stoichiometry: Chemical Calculations

Show all work

Name (Write the name)	Formula (write the formula)	Molecular Mass	How many g, mols or molecules in
iron (III) oxide			mols in 2.50 g?
magnesium acetate			g in 6.55 x 10^{22} of compound?
ammonium sulfate			g in 6.55 mols?
chromium (III) oxide			mols in 2.50 g?
magnesium carbonate			g in 6.55 x 10^{22} of compound?

Ch 3/Worksheet/Empirical and Molecular Formulas   Name: _____

1. The molecular formula of the antifreeze ethylene glycol is $C_2H_6O_2$. What is the empirical formula?

2. Calculate the molecular mass of tetraphosphorus decaoxide, P_4O_{10}, a corrosive substance which can be used as a drying agent.
   a. 469.73 g/mol
   b. 283.89 g/mol
   c. 190.97 g/mol
   d. 139.88 g/mol
   e. 94.97 g/mol

3. Household sugar, sucrose, has the molecular formula $C_{12}H_{22}O_{11}$. What is the % of carbon in sucrose, by mass?
   a. 26.7 %
   b. 33.3 %
   c. 41.4 %
   d. 42.1 %
   e. 52.8 %

4. What is the percent carbon in CH_3CH_2OH?
   a. 13%
   b. 24%
   c. 35%
   d. 46%
   e. 52%

5. Hydroxylamine nitrate contains 29.17 mass % N, 4.20 mass % H, and 66.63 mass % O. Determine its empirical formula.
   a. HNO
   b. H_2NO_2
   c. HN_6O_{16}
   d. $HN_{16}O_7$
   e. H_2NO_3.

6. A well-known reagent in analytical chemistry, dimethylglyoxime, has the empirical formula C_2H_4NO. If its molar mass is 116.1 g/mol, what is the molecular formula of the compound?

7. Nitrogen and oxygen form an extensive series of oxides with the general formula N_xO_y. One of them is a blue solid that comes apart, reversibly, in the gas phase. It contains 36.84% N. What is the empirical formula of this oxide?

8. A sample of indium chloride weighing 0.5000 g is found to contain 0.2404 g of chlorine. What is the empirical formula of the indium compound?

9. What is the average mass, in grams, of one atom of iron ($N_A = 6.022 \times 10^{23}$ mol^{-1})?
   a. 6.02×10^{23} g
   b. 1.66×10^{-24} g
   c. 9.28×10^{-23} g
   d. 55.85 g
   e. 55.85×10^{-23} g

10. The number of hydrogen atoms in 0.050 mol of $C_3H_8O_3$ is
    a. 3.0×10^{22} H atoms
    b. 1.2×10^{23} H atoms
    c. 2.4×10^{23} H atoms
    d. 4.8×10^{23} H atoms
    e. none of these choices is correct

11. What is the mass in grams of 0.250 mol of the common antacid calcium carbonate?
    a. 4.00×10^{2} g
    b. 25.0 g
    c. 17.0 g
    d. 4.00×10^{-2} g
    e. 2.50×10^{-3} g

12. Aluminum oxide, Al_2O_3, is used as a filler for paints and varnishes as well as in the manufacture of electrical insulators. Calculate the number of moles in 47.51 g of Al_2O_3.
    a. 2.377 mol
    b. 2.146 mol
    c. 1.105 mol
    d. 0.4660 mol
    e. 0.4207 mol

Ch 3/ Worksheet/Stoichiometry

Name: _____

Write the equation and balance it where necessary. **Show all work.**

1. How many grams of sodium fluoride (used in water fluoridation and manufacture of insecticides) are needed to form 485 g of sulfur tetrafluoride?

    $$3SCl_2(l) + 4NaF(s) \rightarrow SF_4(g) + S_2Cl_2(l) + 4NaCl(s)$$

    a. 1940 g   b. 1510 g   c. 754 g   d. 205 g   e. 51.3 g

2. What mass, in grams, of sodium carbonate is required for complete reaction with 8.35g of nitric acid to produce sodium nitrate, carbon dioxide, and water?

    a. 28.1 g   b. 14.04 g   c. 4.96 g   d. 7.02 g   e. 400.0 g

3. Ammonia is produced by the reaction: $3 H_2(g) + N_2(g) \rightarrow 2 NH_3(g)$
    a. If $N_2(g)$ is present in excess and 55.6 g of $H_2(g)$ reacts, what is the *theoretical yield* of $NH_3(g)$?
    b. What is the *percent yield* if the actual yield of the reaction is 159 g of $NH_3(g)$?

Chapter 3: Stoichiometry: Chemical Calculations

4. When 61.3 g Cl_2 is reacted with excess, PCl_3 119.3 g PCl_5 is formed. What is the percent yield for the PCl_5 formed? (Hint: calculate the theoretical yield of PCl_5 first.)

$$PCl_3(g) + Cl_2(g) \rightarrow PCl_5(g)$$

   a. 195%   b. 85.0%   c. 66.3%   d. 51.4%   e. 43.7%

5. Write the following equation in symbol form.
Iron (II) sulfide + hydrochloric acid ⟶ iron (II) chloride + dihydrogen sulfide
How much dihydrogen sulfide will be synthesized from 8.50 g iron (II) sulfide? What is the percent yield if 2.60 g of dihydrogen sulfide is actually obtained after the reaction?

Ch 3/ Worksheet/Limiting Reagent   Name: _____

Show all work. Write the equation and balance it where necessary.

1. What is the maximum number of grams of ammonia, NH_3, which can be obtained from the reaction of 10.0 g of H_2 and 80.0 g of N_2? Calculate the grams of excess reagent left.   (Ans: 56.7g, 33.3g)

1. What is the theoretical yield of chromium that can be produced by the reaction of 40.0 g of Cr_2O_3 with 8.00 g of aluminum? Calculate the percent yield if 13.9 g of product is produced. Calculate the grams of excess reagent left.   (15.4g, 90%, 17.5g)

Chapter 4
Reactions in Aqueous Solutions

Electrical Properties of Aqueous Solutions

Electrolytes:
- Arrhenius theory: non, weak, strong (depending on ionization in solution)
- Most ionic compounds are strong electrolytes.
- Some molecular acids are strong electrolytes.
- Most molecular compounds and organic compounds are weak or non electrolytes.

Precipitation Reactions

Solubility rules

Soluble		Insoluble
Group I salts		
Ammonium salts		
Nitrates		
Acetates		
Perchlorates		
Halides	except	Pb^{2+}, Ag^+, Hg_2^{2+}
Sulfates	except	$Sr^{2+}, Ba^{2+}, Ca^{2+}, Pb^{2+}, Hg_2^{2+}, Ag^+$
Except Groups II salts		Carbonates
		Phosphates
Ca^{2+}, Ba^{2+}		Hydroxides
Except Groups II salts		Sulfides

- Writing and predicting products of reactions

Acid base reaction – neutralization

	Acid	+	Base	$\rightarrow$	Salt	+	Water
Eg	HCl	+	NaOH	$\rightarrow$	NaCl	+	H_2O

Reactions of Acids and Bases
- Strong and weak acids: strength depends on concentration of $[H_3O^+]$ in solution.
- Strong and weak bases: strength depends on concentration of $[OH^-]$ in solution.

Oxidation and Reduction

Oxidation	Reduction
• Addition of oxygen	• Removal of oxygen
• Removal of hydrogen	• Addition of hydrogen
• Loss of electrons (LEO)	• Gain of electrons (GER)

Oxidizing and reducing agents

Oxidizing agent
- reduced in the reaction
- causes the other substance to be oxidized
- Oxidation number decreases (as electrons are accepted)
- E.g.: non metals

Reducing agent
- oxidized in the reaction
- causes the other substance to be reduced
- Oxidation number increases (as electrons are given)
- E.g. metals as they give electrons

Oxidation Numbers: charge on any ion

Rules for determining oxidation number:
Neutral species e.g. metals and bimolecular compounds (Cl_2, O_2 etc) have 0
For ions the Oxdn # is the charge on the ion e.g. Cr^{6+} is +6 and SO_4^{2-} is -2
Group I is always +1
Group II is always +2
Fluorine is always -1
Hydrogen is always +1
Oxygen is mostly -2
In group binary compounds VII are -1 (e.g. NaCl, Cl is -1)
  VI are -2 (e.g. K_2S)
  V are -3 (NaN_3)

- Identify the species getting oxidized or reduced in a reaction. (Redox reactions)
- First see if oxygen or hydrogen is gained or lost and then check for electron transfer.
- Balancing redox reactions.

Practical Applications
1. batteries
2. organic chemistry: alcohol
3. industrial processes
4. household chemicals e.g. H_2O_2, benzyl peroxide, bleach
5. food and nutrition: energy from aerobic oxidation of carbohydrate and vitamin C as antioxidant.

Solutions

Solution = solute + solvent (the larger quantity)
Saturated, concentrated and dilute solutions.

Unless specified most solutions will be made in water.

Molar concentration (molarity)
Molarity (M) = mol/L (also the units)

Dilution of solutions
$$M_1V_1 = M_2V_2$$
Initial values = final values
(since number of actual mols of solute does not change only amount of solvent changes)

Solutions in chemical reactions: acid base titration

Chapter 4: Reactions in Aqueous Solutions

Titrations: Quantitative neutralization reactions.
Calculations of titration: e.g. how much acid of a known concentration will it take to neutralize a base of unknown concentration? (One of the chemicals has to have known concentration)
Calculation set up similar to stoichiometry in chapter 3.

$$\left(vol\ Acid \ x\ molarity\ of\ Acid \right) x \left(\frac{mol\ base}{mol\ acid} \right) x\ \frac{1}{vol\ base} = molarity\ of\ base$$

(to calculate mols of acid used)   (from balanced equation)

Chapter 4 Electrolytes and Precipitation Reactions

Aqueous Solutions

- **Solution** - a homogeneous mixture of solute + solvent
  - Solute: the component that is dissolved
  - Solvent: the component that does the dissolving (the larger quantity)
- Aqueous solutions are those in which water is the solvent.
- **Dissociation** - ionic compounds separate into constituent ions when dissolved in solution

$$NaCl(s) \xrightarrow{H_2O} Na^+(aq) + Cl^-(aq)$$

- **Ionization** - formation of ions by molecular compounds when dissolved

$$HCl(g) \xrightarrow{H_2O} H^+(aq) + Cl^-(aq)$$

$$NH_3(g) + H_2O(l) \rightleftharpoons NH_4^+(aq) + OH^-(aq)$$

Notes:

Writing Dissociation Equations

- <u>**Ionic compounds produce ions**</u>:

E.g. calcium chloride:

$$CaCl_2 \rightarrow Ca^{2+} + 2Cl^-$$

1) There is only one mol of calcium, but two mols of chloride ions hence use 2 in front of chloride as coefficient (2Cl$^-$). Don't leave the 2 as a subscript (Cl$_2^-$).
2) Always write the products as ions with the proper valencies – calcium is group two hence 2+.

$$Mg_3(PO_4)_2 \rightarrow 3Mg^{2+} + 2PO_4^{3-}$$

In cases of polyatomic ions – keep them as the polyatomic **ions**, just remove the parenthesis and use the subscript outside as the coefficient in the product/ions.

- <u>**Covalent compounds (except acids) do not ionize:**</u>

CO_2, CH_4 will not ionize.
H_2SO_4 will ionize as follows:

$$H_2SO_4 \rightarrow 2H^+ + SO_4^{2-}$$

Notes:

Chapter 4: Reactions in Aqueous Solutions

Electrolytes

Electrolyte: substance that dissolved in water produces a solution that conducts electricity. Will contain ions.

Strong Electrolyte: substances that dissolve completely in water; 100% dissociation
- All water soluble ionic compounds, strong acids and strong bases

Weak Electrolytes: substances that dissolve partially or dissociate partially in water. This solution does not contain many ions.
- Exist mostly as the molecular form in solution
- Weak acids and weak bases

Nonelectrolyte: substance that dissolved in water produces a solution that does not conduct electricity and does not contain ions.

Notes:

Electrolytes

The best way distinguish between electrolytes is to see if they conduct electricity and with how much intensity a bulb would light up.

Non electrolyte: No conduction of electricity. E.g. distilled water, ethanol (C_2H_5OH), sugar solution

Weak electrolyte: Weak conduction of electricity. E.g. tap water, acetic acid ($HC_2H_3O_2$).

Strong electrolyte: Strong conduction of electricity. E.g. strong acid solutions (HCl), salt (NaCl) solution

© Sapna Gupta

Notes:

Strong Acids

These acids dissociate completely

Acid	Dissociation
Hydrochloric acid	$HCl(aq) \longrightarrow H^+(aq) + Cl^-(aq)$
Hydrobromic acid	$HBr(aq) \longrightarrow H^+(aq) + Br^-(aq)$
Nitric acid	$HNO_3(aq) \longrightarrow H^+(aq) + NO_3^-(aq)$
Chloric acid	$HClO_3(aq) \longrightarrow H^+(aq) + ClO_3^-(aq)$
Sulfuric acid	$H_2SO_4(aq) \longrightarrow 2H^+(aq) + SO_4^{2-}(aq)$

Notes:

Electrolytes – Practical Application

- Our body is about 70% water and we have a number of ionic salts in our body. Some common ions are Na^+, Ca^{2+}, K^+, Cl^-, CO_3^{2-}.
- Electrolytes maintain voltages in our cells; they help with nerve impulses in our nervous system and help with muscle contractions.
- We get electrolytes through our diet and the kidneys are responsible for maintaining an electrolytic balance in the body. If the ions are not in the correct concentration then the above mentioned functions cannot occur.
- One way we lose electrolytes is during sweating. This is two fold – one: we lose water in the body so the concentration of ions change and two: we lose ions also during sweating.
- These ions have to replenished or we can lose muscle control.
- Electrolytes (sports drinks) are commonly used to replace these ions. One has to be careful though – they also have a lot of sugar in them!
- The first electrolyte beverage was invented in University of Florida……

Notes:

Types of Reactions

Two classifications: one how atoms are rearrangement and the other is chemical reaction

1) Atomic Rearrangement
- **Synthesis** (combination): two substances combine to form one.
  $2Na(s) + Cl_2(g) \rightarrow 2NaCl(s)$
- **Double Displacement**: A reaction in which two elements displaces two elements.
  $AgNO_3(aq) + NaCl(aq) \rightarrow AgCl(s) + NaNO_3(aq)$
- **Single displacement**: A reaction where one element displaces one other element.
  $Zn(s) + CuSO_4(aq) \rightarrow ZnSO_4(aq) + Cu(s)$
- **Decomposition**: A reaction in which a single compound reacts to give two or more substances.
  $2HgO(s) \rightarrow 2Hg(l) + O_2(g)$

2) Chemical Classification: Types of Chemical Reactions

Precipitation Reactions: where a solid is formed when two solutions are mixed.

Neutralization Reactions: when an acid and base react to from salt and water.

Oxidation–Reduction Reactions: addition or removal of oxygen and/or transfer of electrons.

Notes:

Precipitation Reactions

- Precipitation (formation of a solid from two aqueous solutions) occurs when product is insoluble in water.
- Reaction type: Double displacement
- What is solubility? Solubility is defined as the maximum amount of a solid that can dissolve in a given amount of solvent at a specified temperature
- Prediction of precipitate is based on solubility rules

Notes:

Solubility Guidelines

Soluble		Insoluble
Group I salts		
Ammonium salts		
Nitrates		
Acetates		
Perchlorates		
Halides	except	Pb^{2+}, Ag^+, Hg_2^{2+}
Sulfates	except	Sr^{2+}, Ba^{2+}, Ca^{2+}, Pb^{2+}, Hg_2^{2+}, Ag^+
Except Groups II salts		Carbonates
		Phosphates
Ca^{2+}, Ba^{2+}		Hydroxides
Except Groups II salts		Sulfides

Notes:

Solved Problems

1) Identify the Precipitate

$$Pb(NO_3)_2(aq) + 2NaI(aq) \rightarrow 2NaNO_3 + PbI_2$$

PbI_2 – according to solubility rules

2) Classify the following as soluble or insoluble in water
- $Ba(NO_3)_2$ soluble
- AgI insoluble
- $Mg(OH)_2$ insoluble

Notes:

Chapter 4: Reactions in Aqueous Solutions

Writing Equations in Aqueous Solutions

A chemical equation in which the reactants and products are written as if they were molecular substances, even though they may actually exist in solution as ions.

Symbols indicating the states are include: (s), (l), (g), (aq).

For example:

Molecular Equation:

$$AgNO_3(aq) + NaCl(aq) \rightarrow AgCl(s) + NaNO_3(aq)$$

Although $AgNO_3$, NaCl, and $NaNO_3$ exist as ions in aqueous solutions, they are written as compounds in the molecular equation.

Ionic Equation:

$$Ag^+(aq) + NO_3^-(aq) + Na^+(aq) + Cl^-(aq) \rightarrow AgCl(s) + Na^+(aq) + NO_3^-(aq)$$

All compounds that dissociate are shown as ions.

Net Ionic Equation:

In this the **spectator ions** (ions on both sides of the equation) are eliminated.

$$Ag^+(aq) + NO_3^-(aq) + Na^+(aq) + Cl^-(aq) \rightarrow AgCl(s) + Na^+(aq) + NO_3^-(aq)$$

Net ionic equation represents the ions reacting. Those will be (g), (l) and (s) products formed.

$$Ag^+(aq) + Cl^-(aq) \rightarrow AgCl(s)$$

Notes:

Solved Problems

Decide whether the following reaction occurs. If it does, write the molecular, ionic, and net ionic equations.

$KBr + MgSO_4 \rightarrow$

Determine the product formulas by double displacement method
- K^+ and SO_4^{2-} make K_2SO_4
- Mg^{2+} and Br^- make $MgBr_2$

Determine whether the products are soluble:
K_2SO_4 is soluble and $MgBr_2$ is soluble

$$KBr + MgSO_4 \rightarrow \text{no reaction}$$

Notes:

Chapter 4: Reactions in Aqueous Solutions

Solved Problems

Decide whether the following reaction occurs. If it does, write the molecular, ionic, and net ionic equations.

NaOH + MgCl$_2$ →

Determine the product formulas by double displacement method
- Na$^+$ and Cl$^-$ make NaCl
- Mg^{2+} and OH$^-$ make Mg(OH)$_2$

Determine whether the products are soluble
- NaCl is soluble and Mg(OH)$_2$ is insoluble

Molecular Equation
Balance the reaction and include state symbols

2NaOH(aq) + MgCl$_2$(aq) → 2NaCl(aq) + Mg(OH)$_2$(s)

Ionic Equation

2Na$^+$(aq) + 2OH$^-$(aq) + Mg^{2+}(aq) + 2Cl$^-$(aq) → 2Na$^+$(aq) + 2Cl$^-$(aq) + Mg(OH)$_2$(s)

Net Ionic Equation

2OH$^-$(aq) + Mg^{2+}(aq) → Mg(OH)$_2$(s)

Notes:

One more....

Decide whether the following reaction occurs. If it does, write the molecular, ionic, and net ionic equations.

K$_3$PO$_4$ + CaCl$_2$ →

Determine the product formulas:
- K$^+$ and Cl$^-$ make KCl; Ca^{2+} and PO$_4^{3-}$ make Ca$_3$(PO$_4$)$_2$

Determine whether the products are soluble:
- KCl is soluble and Ca$_3$(PO$_4$)$_2$ is insoluble

Molecular Equation
(Balance the reaction and include state symbols)

2K$_3$PO$_4$(aq) + 3CaCl$_2$(aq) → 6KCl(aq) + Ca$_3$(PO$_4$)$_2$(s)

Ionic Equation

6K$^+$(aq) + 2PO$_4^{3-}$(aq) + 3Ca^{2+}(aq) + 6Cl$^-$(aq) → 6K$^+$(aq) + 6Cl$^-$(aq) + Ca$_3$(PO$_4$)$_2$(s)

Net Ionic Equation

2PO$_4^{3-}$(aq) + 3Ca^{2+}(aq) → Ca$_3$(PO$_4$)$_2$(s)

Notes:

Another one....

Aqueous solutions of silver nitrate and sodium sulfate are mixed. Write the net ionic reaction.

$$2AgNO_3(aq) + Na_2SO_4(aq) \rightarrow 2NaNO_3(?) + Ag_2SO_4(?)$$

Determine solubility of salts. All nitrates are soluble but silver sulfate is insoluble

Molecular Equation

$$2AgNO_3(aq) + Na_2SO_4(aq) \rightarrow 2NaNO_3(aq) + Ag_2SO_4(s)$$

Ionic equation

$$2Ag^+(aq) + 2NO_3^-(aq) + 2Na^+(aq) + SO_4^{2-}(aq) \rightarrow 2Na^+(aq) + 2NO_3^-(aq) + Ag_2SO_4(s)$$

Cancel spectators

$$2Ag^+(aq) + 2\cancel{NO_3^-}(aq) + 2\cancel{Na^+}(aq) + SO_4^{2-}(aq) \rightarrow 2\cancel{Na^+}(aq) + 2\cancel{NO_3^-}(aq) + Ag_2SO_4(s)$$

Net ionic equation

$$2Ag^+(aq) + SO_4^{2-}(aq) \rightarrow Ag_2SO_4(s)$$

Key Words and Concepts

- **Ions in Aqueous Solution**
  - Electrolytes
  - Acids
- **Types of Chemical Reactions**
  - Synthesis
  - Double displacement
  - Single displacement
  - Decomposition
- **Precipitation Reactions**
  - Solubility Rules
  - Molecular, Ionic and Net Ionic Equations

Ch 4/PowerPoint Study-1 Electrolytes, Precipitation Reactions

Name: _____

Answer these questions as you are watching the videos. They are due in class.
These questions are not just for you to answer but also to prepare you for the exam.
<u>*Make sure you understand what you are writing and not just copy from the text book.*</u> **Show all work.**

1. Circle the compounds below that will dissociate to give ions in water. (Hint: think which one will give ions in solution. Ionic compound ionize – what about HCl?)

   $NaNO_3$ CH_4 HCl   KCl

2. For the compounds above, write which ones will be an electrolyte.

3. Write the equation that shows dissociation and the ions produced in the following compounds.
   a. $CaS_{(aq)} \longrightarrow$

   b. $Na_2SO_{4(aq)} \longrightarrow$

   c. $Al_2(SO_4)_{3(aq)} \longrightarrow$

4. Look at the solubility guidelines and predict if the following compounds will be soluble in water.

   $PbSO_4$ KCl   CaS   AgCl $NaNO_3$ $Mg(OH)_2$

5. Write the molecular, ionic and net ionic equation for the reaction between copper (II) sulfate and barium chloride.

   Molecular:

   Ionic:

   Net Ionic:

Chapter 4 Electrolytes Acid-Base (Neutralization) Oxidation-Reduction (Redox) Reactions

Dr. Sapna Gupta

Types of Reactions

Two classifications: one how atoms are rearrangement and the other is chemical reaction

1) Atomic Rearrangement
- **Synthesis** (combination): two substances combine to form one.
  $2Na(s) + Cl_2(g) \rightarrow 2NaCl(s)$
- **Double Displacement**: A reaction in which two elements displaces two elements.
  $AgNO_3(aq) + NaCl(aq) \rightarrow AgCl(s) + NaNO_3(aq)$
- **Single displacement**: A reaction where one element displaces one other element.
  $Zn(s) + CuSO_4(aq) \rightarrow ZnSO_4(aq) + Cu(s)$
- **Decomposition**: A reaction in which a single compound reacts to give two or more substances.
  $2HgO(s) \rightarrow 2Hg(l) + O_2(g)$

2) Chemical Classification: Types of Chemical Reactions

Precipitation Reactions: where a solid is formed when two solutions are mixed.
Neutralization Reactions: when an acid and base react to from salt and water.
Oxidation–Reduction Reactions: addition or removal of oxygen and/or transfer of electrons.

Strong Acids

These acids dissociate completely

Hydrochloric acid	$HCl(aq)$	$H^+(aq) + Cl^-(aq)$
Hydrobromic acid	$HBr(aq)$	$H^+(aq) + Br^-(aq)$
Nitric acid	$HNO_3(aq)$	$H^+(aq) + NO_3^-(aq)$
Chloric acid	$HClO_3(aq)$	$H^+(aq) + ClO_3^-(aq)$
Sulfuric acid	$H_2SO_4(aq)$	$2H^+(aq) + SO_4^{2-}(aq)$

Chapter 4: Reactions in Aqueous Solutions

Neutralization Reactions (acid-base)

Acids	Bases
Arrhenius Acid A substance that produces hydrogen ions, H^+, when dissolved in water.	**Arrhenius Base** A substance that produces hydroxide ions, OH^-, when dissolved in water.
Brønsted–Lowry Acid The species (molecule or ion) that donates a proton, H^+, to another species in a proton–transfer reaction.	**Brønsted–Lowry Base** The species (molecule or ion) that accepts a proton, H^+, from another species in a proton–transfer reaction.
Sour	Bitter
Corrosive	Caustic, slippery
pH value 1-7	pH value 7-14
Strong acids (inorganic acids) – ionize completely in water, e.g.: HNO_3, H_2SO_4, $HClO_4$, HCl, HBr, HI	Strong bases (inorganic bases) – ionize completely in water; most are hydroxides, e.g.: $NaOH$, KOH, $Ca(OH)_2$
Weak acids – ionize partially in water, e.g. HF Organic acid: $HC_2H_3O_2$ (CH_3COOH)	Weak bases– ionize partially in water, e.g.: NH_4OH, Na_2CO_3, $NaHCO_3$ organic bases: CH_3NH_2

Notes:

More on Acids-Bases

Indicators: these are chemicals used to determine if an acid or base is strong or weak by changing colors. Below are the colors for universal indicator.

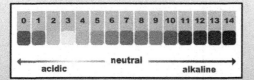

Polyprotic Acid: An acid that results in two or more acidic hydrogens per molecule. E.g. HCl has only one proton to give; but H_2SO_4, sulfuric acid can give 2 protons.

Notes:

Chapter 4: Reactions in Aqueous Solutions

Acid-Base Neutralization Reactions

Neutralization Reaction:
- Almost all acid base reactions are double displacement reactions.
- Most will produce a salt and water as product.
- Carbonates and sulfites give CO_2 and SO_2 gases in product.

Neutralization: Reaction between an acid and a base

$$\text{Acid} + \text{Base} \rightarrow \text{Salt} + \text{Water}$$

Molecular equation:
$$HCl(aq) + NaOH(aq) \rightarrow NaCl(aq) + H_2O(l)$$

Ionic equation:
$$H^+(aq) + \cancel{Cl^-}(aq) + \cancel{Na^+}(aq) + OH^-(aq) \rightarrow \cancel{Na^+}(aq) + \cancel{Cl^-}(aq) + H_2O(l)$$

Net ionic equation:
$$H^+(aq) + OH^-(aq) \rightarrow H_2O(l)$$

Notes:

Neutralization Reactions

Write the molecular, ionic, and net ionic equations for the neutralization of sulfurous acid, H_2SO_3, by potassium hydroxide, KOH

The reaction is a double displacement reaction.

Molecular Equation
(Balance the reaction and include state symbols)
$$H_2SO_3(aq) + 2KOH(aq) \rightarrow 2H_2O(l) + K_2SO_3(aq)$$

Ionic Equation
$$H_2SO_3(aq) + 2\cancel{K^+}(aq) + 2OH^-(aq) \rightarrow 2H_2O(l) + 2\cancel{K^+}(aq) + SO_3^{2-}(aq)$$

Net Ionic Equation
$$H_2SO_3(aq) + 2OH^-(aq) \rightarrow 2H_2O(l) + SO_3^{2-}(aq)$$

Notes:

Neutralization Reactions Producing Gases

Sulfides, carbonates, sulfites react with acid to form a gas.

$Na_2S(aq) + 2HCl(aq) \rightarrow 2NaCl(aq) + H_2S(g)$

$Na_2CO_3(aq) + 2HCl(aq) \rightarrow 2NaCl(aq) + H_2O(l) + CO_2(g)$

$Na_2SO_3(aq) + 2HCl(aq) \rightarrow 2NaCl(aq) + H_2O(l) + SO_2(g)$

Baking soda (sodium hydrogen carbonate) reacting with acetic acid in vinegar to give bubbles of carbon dioxide.

Notes:

Neutralization Reaction – another one

Molecular Equation
(Balance the reaction and include state symbols)
$CuCO_3(s) + 2HCl(aq) \rightarrow CuCl_2(aq) + H_2O(l) + CO_2(g)$

Ionic Equation
$CuCO_3(s) + 2H^+(aq) + 2Cl^-(aq) \rightarrow Cu^{2+}(aq) + 2Cl^-(aq) + H_2O(l) + CO_2(g)$

Net Ionic Equation
$CuCO_3(s) + 2H^+(aq) \rightarrow Cu^{2+}(aq) + H_2O(l) + CO_2(g)$

Notes:

Redox Reactions

Oxidation	Reduction
Addition of oxygen	Removal of oxygen
Removal of hydrogen	Addition of hydrogen
Loss of electrons (LEO)	Gain of electrons (GER)
Metals lose electrons hence undergo oxidation	Non metals gain electrons hence undergo reduction
Reducing agents – something that causes reduction of another element and gets oxidized (loses electrons) itself	Oxidizing agent – an element that causes oxidation of another element and gets reduced (gains electrons) itself

Writing Redox Reactions

Example
$$Mg(s) + CuSO_4(aq) \rightarrow MgSO_4(aq) + Cu(s)$$

gaining 2 electrons, reduction

$$Mg(s) + Cu^{2+}(aq) \rightarrow Mg^{2+}(aq) + Cu(s)$$

loses 2 electrons, oxidation

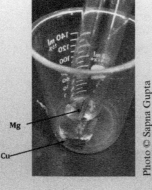

Half Reactions:
Zinc is losing 2 electrons and oxidized.
It is the reducing agent
$$Mg(s) \rightarrow Mg^{2+}(aq) + 2e^-$$
Copper ions are gaining the 2 electrons.
It is the oxidizing agent
$$Cu^{2+}(aq) + 2e^- \rightarrow Cu(s)$$

Chapter 4: Reactions in Aqueous Solutions

Rules for Assigning Oxidation Numbers

1. **Elements:** The oxidation number of an element is zero.
2. **Monatomic ions:** The oxidation number of a monatomic ion equals the charge on the ion.
3. **Oxygen:** The oxidation number of oxygen is –2 in most compounds. (An exception is O in H_2O_2 and other peroxides, where the oxidation number is –1.)
4. **Hydrogen:** The oxidation number of hydrogen is +1 in most of its compounds. (The oxidation number of hydrogen is –1 in binary compounds with a metal such as CaH_2.)
5. **Halogens:** The oxidation number of fluorine is –1. Each of the other halogens (Cl, Br, I) has an oxidation number of –1 in binary compounds, except when the other element is another halogen above it in the periodic table or the other element is oxygen.
6. **Compounds and ions:** The sum of the oxidation numbers of a compound is *zero*. The sum of the oxidation numbers of the atoms in a polyatomic ion equals the charge on the ion.

Notes:

Oxidation Numbers on the Periodic Table

IA	IIA											IIIA	IVA	VA	VIA	VIIA	VIIIA
1 H +1, –1																	2 He
3 Li +1	4 Be +2											5 B +3	6 C +4, +2, –4	7 N +5, +4, +3, +2, +1, –3	8 O +2, –1/2, –1, –2	9 F –1	10 Ne
11 Na +1	12 Mg +2											13 Al +3	14 Si +4, –4	15 P +5, +3, –3	16 S +6, +4, +2, –2	17 Cl +7, +6, +5, +4, +3, +1, –1	18 Ar
		IIIB	IVB	VB	VIB	VIIB	VIIIB	VIIIB	VIIIB	IB	IIB						
19 K +1	20 Ca +2	21 Sc +3	22 Ti +4, +3	23 V +5, +4, +3, +2	24 Cr +6, +3, +2	25 Mn +7, +6, +4, +3, +2	26 Fe +3, +2	27 Co +3, +2	28 Ni +2	29 Cu +2, +1	30 Zn +2	31 Ga +3	32 Ge +4, –4	33 As +5, +3, –3	34 Se +6, +4, –2	35 Br +5, +1, –1	36 Kr
37 Rb +1	38 Sr +2	39 Y +3	40 Zr +4	41 Nb +5, +4	42 Mo +6, +4, +3	43 Tc +7, +6, +4	44 Ru +8, +6, +4, +3	45 Rh +4, +3, +2	46 Pd +4, +2	47 Ag +1	48 Cd +2	49 In +3	50 Sn +4, +2	51 Sb +5, +3, –3	52 Te +6, +4, –2	53 I +7, +5, +1, –1	54 Xe
55 Cs +1	56 Ba +2	57 La +3	72 Hf +4	73 Ta +5	74 W +6, +4	75 Re +7, +6, +4	76 Os +8, +4	77 Ir +4, +3	78 Pt +4, +2	79 Au +3, +1	80 Hg +2, +1	81 Tl +3, +1	82 Pb +4, +2	83 Bi +5, +3	84 Po +2	85 At –1	86 Rn

Notes:

Activity Series

Loses electrons easily

Element		Oxidation Half Reaction
Lithium	Li →	$Li^+ + 1e^-$
Potassium	K →	$K^+ + 1e^-$
Barium	Ba →	$Ba^{+2} + 2e^-$
Calcium	Ca →	$Ca^{+2} + 2e^-$
Sodium	Na →	$Na^+ + 2e^-$
Magnesium	Mg →	$Mg^{+2} + 2e^-$
Zinc	Zn →	$Zn^{+2} + 2e^-$
Chromium	Cr →	$Cr^{+3} + 3e^-$
Iron	Fe →	$Fe^{+2} + 2e^-$
Cadmium	Cd →	$Cd^{+2} + 2e^-$
Cobalt	Co →	$Co^{+2} + 2e^-$
Nickel	Ni →	$Ni^{+2} + 2e^-$
Tin	Sn →	$Sn^{+2} + 2e^-$
Lead	Pb →	$Pb^{+2} + 2e^-$
Hydrogen	H →	$H^+ + 1e^-$
Copper	Cu →	$Cu^{+2} + 2e^-$
Silver	Ag →	$Ag^+ + 1e^-$
Mercury	Hg →	$Hg^{+2} + 2e^-$
Platinum	Pt →	$Pt^{+2} + 2e^-$
Gold	Au →	$Au^{+2} + 2e^-$

Does not lose electrons easily

Notes:

Assigning Oxidation States

Assign oxidation numbers for all elements in each species

1) $MgBr_2$: Mg +2, Br −1 x 2 = −2; *+2 +(−2)* = total charge of 0
2) ClO_2^- : O −2 x 2 = −4; Cl +3; *(−4) + (+3)* = −1 (charge left over on ion)
3) Assign oxidation number of Mn in K**Mn**O_4

K	Mn	O	
1(+1)	+ 1(oxidation number of Mn)	+ 4(−2)	= 0
1	+ 1(oxidation number of Mn)	+ (−8)	= 0
(−7) + (oxidation number of Mn)			= 0

Oxidation number of Mn = +7

4) What is the oxidation number of Cr in dichromate, **Cr**$_2O_7^{2-}$?

Cr	O	
2(oxidation number of Cr)	+ 7(−2)	= −2
2(oxidation number of Cr)	+ (−14)	= −2
2(oxidation number of Cr)		= +12

Oxidation number of Cr = +6

Notes:

Chapter 4: Reactions in Aqueous Solutions

Balancing Redox Equations (electronically)

Balance the following reaction.
$$Zn(s) + Ag^+(aq) \rightarrow Zn^{2+}(aq) + Ag(s)$$

Oxidation Numbers   0   +   +2   0

Next, write the unbalanced half-reactions.

$Zn(s) \rightarrow Zn^{2+}(aq)$ (oxidation)
$Ag^+(aq) \rightarrow Ag(s)$ (reduction)

Now, balance the charge in each half reaction by adding electrons.

$Zn(s) \rightarrow Zn^{2+}(aq) + 2e^-$ (oxidation)
$e^- + Ag^+(aq) \rightarrow Ag(s)$ (reduction)

Each half-reaction should have the same number of electrons. To do this, multiply each half-reaction by a factor so that when the half-reactions are added, the electrons cancel.

$Zn(s) \rightarrow Zn^{2+}(aq) + 2e^-$ (oxidation)
$2e^- + 2Ag^+(aq) \rightarrow 2Ag(s)$ (reduction)

Lastly, add the two half-reactions together.

$$Zn(s) + 2Ag^+(aq) \rightarrow Zn^{2+}(aq) + 2Ag(s)$$

Notes:

One more Balancing Redox Equation

$$FeI_3(aq) + Mg(s) \rightarrow Fe(s) + MgI_2(aq)$$

The oxidation numbers are given below the reaction.

$FeI_3(aq) + Mg(s) \rightarrow Fe(s) + MgI_2(s)$
+3 −1   0   0   +2 −1

Now, write the half-reactions. Since Iodide is a spectator ion it is omitted at this point.

$Mg(s) \rightarrow Mg^{2+}(aq)$ (oxidation)
$Fe^{3+}(aq) \rightarrow Fe(s)$ (reduction)

Balancing the half-reactions:

$Mg(s) \rightarrow Mg^{2+}(aq) + 2e^-$ (oxidation)
$Fe^{3+}(aq) + 3e^- \rightarrow Fe(s)$ (reduction)

Multiply the oxidation half-reaction by 3 and the reduction half-reaction by 2.

$3Mg(s) \rightarrow 3Mg^{2+}(aq) + 6e^-$ (oxidation)
$2Fe^{3+}(aq) + 6e^- \rightarrow 2Fe(s)$ (reduction)

Add the half-reactions together.

$2Fe^{3+}(aq) + 3Mg(s) \rightarrow 2Fe(s) + 3Mg^{2+}(aq)$

Now, return the spectator ion, I^-.

$2FeI_3(aq) + 3Mg(s) \rightarrow 2Fe(s) + 3MgI_2(aq)$

Notes:

Chapter 4: Reactions in Aqueous Solutions

Types of Redox Reactions

Displacement reactions

A common reaction: active metal replaces (displaces) a metal ion from a solution (use the activity series to predict if reaction will take place)

$$Mg(s) + CuCl_2(aq) \rightarrow Cu(s) + MgCl_2(aq)$$

Decomposition reactions

$$2KClO_3(s) \rightarrow 2KCl(s) + 3O_2(g)$$

Combination Reactions

$$2H_2(g) + O_2(g) \rightarrow 2H_2O(l)$$

Combustion reactions

Common example, hydrocarbon fuel reacts with oxygen to produce carbon dioxide and water

$$CH_4(g) + O_2(g) \rightarrow H_2O(l) + CO_2(g)$$

Notes:

Solved Example

Classify the following reactions as precipitation; acid-base; or redox reaction and any other classification that can describe the reaction.

$2H_2 + O_2 \rightarrow 2H_2O$
    Redox (combustion, combination)

$Zn + H_2SO_4 \rightarrow ZnSO_4 + H_2$
    Redox (single displacement)

$H_2O + NH_3 \rightarrow NH_4^+ + OH^-$
    Acid-base (double displacement)

$6FeSO_4 + K_2Cr_2O_7 + 7H_2SO_4 \rightarrow Cr_2(SO_4)_3 + 3Fe_2(SO_4)_3 + K_2SO_4 + 7H_2O$
    Redox

$2NaCl + Pb(NO_3)_2 \rightarrow PbCl_2 + 3NaNO_3$
    Precipitation (double displacement)

Notes:

Chapter 4: Reactions in Aqueous Solutions

Key Words and Concepts

- **Types of Chemical Reactions**
  - Acid–Base Reactions
  - Oxidation–Reduction Reactions
    - Oxidation
    - Reduction
    - Reducing agent
    - Oxidizing agent
    - Half reactions
    - Activity series

Notes:

Ch 4/PowerPoint Study-2 Acid Base and Redox Reactions

Name: _____

Answer these questions as you are watching the videos. They are due in class.
These questions are not just for you to answer but also to prepare you for the exam.
Make sure you understand what you are writing and not just copy from the text book. **Show all work.**

1. Which of the following compounds are acids and which are bases? And which is neither?

    NaOH   NaNO$_3$ H$_3$PO$_4$ NH$_4$OH   HNO$_3$

2. Write the molecular, ionic and net ionic equation for the following reaction. sulfuric acid and aluminum hydroxide.

3. Calculate the oxidation number on sulfur for each of the following compounds.
    a. H$_2$S   b. SO$_2$ c. SCl$_2$ d. H$_2$SO$_3$ e. Na$_2$SO$_4$

4. Write the chemical equation for the reaction of aluminum and oxygen (redox reaction). Write the oxidation numbers on each of elements on reactant and products.

Chapter 4 Solution Concentration

Dr. Sapna Gupta

Concentrations of Solutions

- A solution is solute dissolved in a solvent.
- To quantify and know exactly how much of a solute is present in a certain amount of solvent, one will need to calculate concentrations.
- Concentrations are given in
  - percent solutions
    - mass/mass % - (g of solute/g of solution) x 100%
    - mass/volume % - (g of solute/mL of solution) x 100%
    - volume/volume % - (mL of solute/mL of solution) x 100%
  - Molarity (mol/L of solution)- used more in Chemistry
  - Molality (mol/Kg of solution) – this is used more in Biology

Notes:

Molar Concentration, Molarity (M)

In this chapter we will study **Molarity** – which is moles in a L of solution.
- Molarity is represented by M and the formula is given below

$$\text{Molarity} = \frac{\text{moles of solute}}{\text{L of solution}}$$

- Moles are converted to grams in order to make the solution in the lab.
- To prepare a solution, add the measured amount of solute to a volumetric flask, then add water to bring the solution to the mark on the flask.

Weigh the solute → add to volumetric flask and add water

- A 3M solution of NaCl means there are 3 moles of NaCl in the solution.
- If you have a 200 mL of 2 M HCl – that means that there are 2 mols of HCl in 1 L solution. If you want to know how many grams of HCl you have in 200 mL then you will have to calculate the amount of moles in 200 mL of that solution using the Molarity equation; then you can calculate the grams from those moles.

Notes:

Chapter 4: Reactions in Aqueous Solutions

Example: You place a 1.52-g of potassium dichromate, $K_2Cr_2O_7$, into a 50.0-mL volumetric flask. You then add water to bring the solution up to the mark on the neck of the flask. What is the molarity of $K_2Cr_2O_7$ in the solution?

Molar mass of $K_2Cr_2O_7$ is 294 g/mol

$$\frac{1.52 \text{ g} \frac{1 \text{ mol}}{294 \text{ g}}}{50.0 \times 10^{-3} \text{ L}} = 0.103 \, M$$

Example: A solution of sodium chloride used for intravenous transfusion (physiological saline solution) has a concentration of 0.154 M NaCl. How many moles of NaCl are contained in 500.-mL of physiological saline? How many grams of NaCl are in the 500.-mL of solution?

$M = \frac{mol}{L}$

mol = $M \cdot L$
= 0.154 M • 0.500 L
= 0.0770 mol NaCl

Molar mass NaCl = 58.4 g

0.0770 mol $\frac{58.4 \text{ g}}{1 \text{ mol}}$
= 4.50 g NaCl

Notes:

Example: Calculate the molarity of a solution prepared by dissolving 45.00 grams of KI into a total volume of 500.0 mL.

$$\frac{45.00 \text{ g KI}}{500.0 \text{ mL}} \times \frac{1 \text{ mol KI}}{166.0 \text{ g KI}} \times \frac{1000 \text{ mL}}{1 \text{ L}} = 0.5422 \, M$$

Example: How many milliliters of 3.50 M NaOH can be prepared from 75.00 grams of the solid?

$$75.00 \text{ g NaOH} \times \frac{1 \text{ mol NaOH}}{40.00 \text{ g NaOH}} \times \frac{1 \text{ L}}{3.50 \text{ mol NaOH}} \times \frac{1000 \text{ mL}}{1 \text{ L}} = 536 \text{ mL}$$

Notes:

Chapter 4: Reactions in Aqueous Solutions — 139

Dilution

- When a higher concentration solution is used to make a less-concentration solution, the moles of solute are determined by the amount of the higher-concentration solution.
- The number of moles of solute remains constant when more water is added.

$$M_i V_i = M_f V_f$$

Note:
The units on V_i and V_f must match.

Diluting a solution quantitatively requires specific glassware.

A volumetric flask, shown here, is used in dilution.

Note: the number of molecules has not changed. Only the volume of solvent is increased.

Source: Sapna Gupta

Example: A saturated stock solution of NaCl is 6.00 M. How much of this stock solution is needed to prepare 1.00-L of physiological saline solution (0.154 M)?

$$M_i V_i = M_f V_f$$
$$V_i = \frac{M_f V_f}{M_i}$$

$$V_i = \frac{(0.154\ M)(1.00\ L)}{6.00\ M}$$

$$V_i = 0.0257\ L\ \text{or}\ 25.7\ mL$$

Example: For the next experiment the class will need 250. mL of 0.10 M CuCl$_2$. There is a bottle of 2.0 M CuCl$_2$. Describe how to prepare this solution. How much of the 2.0 M solution do we need?

Concentrated: 2.0 M use ? mL (V_c)
Diluted: 250. mL of 0.10 M

$$M_c V_c = M_d V_d$$
$$(2.0\ M)(V_c) = (0.10\ M)(250.\ mL)$$
$$V_c = 12.5\ mL$$

12.5 mL of the concentrated solution are needed; add enough distilled water to prepare 250. mL of the solution.

Chapter 4: Reactions in Aqueous Solutions

Key Words/Concepts

- Solutions
  - Solvent
  - Solute
- Molarity (mol/L)
- Dilutions ($M_i V_i = M_f V_f$)

Notes:

Ch 4/ PowerPoint Study-3 Solution Concentration   Name: _____

Answer these questions as you are watching the videos. They are due in class.
These questions are not just for you to answer but also to prepare you for the exam.
Make sure you understand what you are writing and not just copy from the text book. **Show all work.**

1. Calculate the molarity of a solution that has 0.936 mol in 450 mL.

2. What is the molarity of a solution that is made from 24.0 g of NaCl dissolved in 250 mL of water? (Strategy: calculate mols of NaCl first then divide by L of water.)

3. How many grams of sodium sulfate are in 200 mL of a 0.250 M solution? (Strategy: a) how many mols are in 200 mL of water? b) convert the mols to grams of sodium sulfate)

4. Calculate the new concentration when 0.020 L of a 4.50 M sulfuric acid solution is diluted to 2.00 L?

Chapter 4 Solution Stoichiometry

Dr. Sapna Gupta

Solution Stoichiometry

- In solution stoichiometry you have to presume that soluble ionic compounds dissociate completely in solution.
- Then using mole ratios we can calculate the concentration of all species in solution.
- There are three common types of stoichiometric calculations
  - **Quantitative Analysis:** The determination of the amount of a substance or species present in a material.
  - **Volumetric Analysis:** A type of quantitative analysis based on titration.
  - **Gravimetric Analysis:** A type of quantitative analysis in which the amount of a species in a material is determined by converting the species to a product that can be isolated completely and weighed.

Notes:

Volumetric Analysis - Titrations

A procedure for determining the amount of substance A by adding a carefully measured volume with a known concentration of B until the reaction of A and B is just complete. This can be for precipitation, neutralization or redox reactions.

- **Standardization** is the determination of the exact concentration of a solution.
- **Equivalence point** represents completion of the reaction.
- **Endpoint** is where the titration is stopped.
- An **indicator** is used to signal the endpoint.

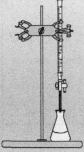

Source: Sapna Gupta

Notes:

Chapter 4: Reactions in Aqueous Solutions

Gravimetric Analysis

- In gravimetric analysis precipitation reactions are carried out.
- After the reaction the product is precipitated and collected in a crucible or filter paper.
- The precipitate is weighed and then using mole ratios we can calculate the concentration of all species in original solution.

The beaker below shows the forming a precipitate.

The precipitate is filtered. It can then be dried and weighed. Then concentration of the desired ions can be calculated

Source: Sapna Gupta

Notes:

Example: Find the concentration of all species in a 0.25 M solution of $MgCl_2$

$MgCl_2 \rightarrow Mg^{2+} + 2Cl^-$

Given: $MgCl_2$ = 0.25 M

$[Mg^{2+}]$ = 0.25 M (1:1 ratio)

$[Cl^-]$ = 0.50 M (1:2 ratio)

Example: A soluble silver compound was analyzed for the percentage of silver by adding sodium chloride solution to precipitate the silver ion as silver chloride. If 1.583 g of silver compound gave 1.788 g of silver chloride, what is the mass percent of silver in the compound?

Strategy: g AgCl → mol AgCl → mol Ag → g Ag → % Ag

Molar mass of silver chloride (AgCl) = 143.32 g

$$1.788 \text{ g AgCl} \times \frac{1 \text{ mol AgCl}}{143.32 \text{ g AgCl}} \times \frac{1 \text{ mol Ag}}{1 \text{ mol AgCl}} \times \frac{107.9 \text{ g Ag}}{1 \text{ mol Ag}} = 1.346 \text{ g Ag in the compound}$$

$$\frac{1.346 \text{ g Ag}}{1.583 \text{ g silver compound}} \times 100\% = 85.03\% \text{ Ag}$$

Notes:

Chapter 4: Reactions in Aqueous Solutions — 145

Example: Zinc sulfide reacts with hydrochloric acid to produce hydrogen sulfide gas:

$$ZnS(s) + 2HCl(aq) \rightarrow ZnCl_2(aq) + H_2S(g)$$

How many milliliters of 0.0512 M HCl are required to react with 0.392 g ZnS?

Strategy: g ZnS → mol ZnS → mol HCl (mol ratio from eq) → vol HCl (using Molarity)

Molar mass of ZnS = 97.45 g

$$0.392 \text{ g ZnS} \times \frac{1 \text{ mol ZnS}}{97.45 \text{ g ZnS}} \times \frac{2 \text{ mol HCl}}{1 \text{ mol ZnS}} \times \frac{1 \text{ L solution}}{0.0512 \text{ mol HCl}}$$

$$= 0.157 \text{ L} = 157 \text{ mL HCl solution}$$

Notes:

Example: A dilute solution of hydrogen peroxide is sold in drugstores as a mild antiseptic. A typical solution was analyzed for the percentage of hydrogen peroxide by titrating it with potassium permanganate:

$$5H_2O_2(aq) + 2KMnO_4(aq) + 6H^+(aq) \rightarrow 8H_2O(l) + 5O_2(g) + 2K^+(aq) + 2Mn^{2+}(aq)$$

What is the mass percent of H_2O_2 in a solution if 57.5 g of solution required 38.9 mL of 0.534 M KMnO$_4$ for its titration?

Strategy: mols KMnO$_4$ → mols H_2O_2 → mass H_2O_2 → % H_2O_2

Molar mass of H_2O_2 = 34.01 g

$$38.9 \times 10^{-3} \text{ L} \times \frac{0.534 \text{ mol KMnO}_4}{1 \text{ L}} \times \frac{5 \text{ mol H}_2\text{O}_2}{2 \text{ mol KMnO}_4} \times \frac{34.01 \text{ g H}_2\text{O}_2}{1 \text{ mol H}_2\text{O}_2} = 1.77 \text{ g H}_2\text{O}_2$$

$$\frac{1.77 \text{ g H}_2\text{O}_2}{57.5 \text{ g solution}} \times 100\% = 3.07\% \text{ H}_2\text{O}_2$$

Notes:

Chapter 4: Reactions in Aqueous Solutions

Example: A student measured exactly 15.0 mL of an unknown monoprotic acidic solution and placed in an Erlenmeyer flask. An indicator was added to the flask. At the end of the titration the student had used 35.0 mL of 0.12 M NaOH to neutralize the acid. Calculate the molarity of the acid.

Strategy: mols NaOH → mols acid (from eq) → Molarity of acid

$$0.035 \text{ L NaOH} \times \frac{0.12 \text{ mol NaOH}}{1 \text{ L}} \times \frac{1 \text{ mol acid}}{1 \text{ mol base}} = 0.0042 \text{ mol acid}$$

$$M = \frac{0.0042 \text{ mol}}{0.015 \text{ L}} = 0.28 \text{ M acid}$$

Example: Calculate the molarity of 25.0 mL of a monoprotic acid if it took 45.50 mL of 0.25 M KOH to neutralize the acid.

$$\frac{0.25 \text{ mol KOH}}{\text{L}} \times 0.04550 \text{ L} \times \frac{1 \text{ mol acid}}{1 \text{ mol KOH}} = 0.01338 \text{ mol acid}$$

$$\frac{0.01338 \text{ mol acid}}{0.0250 \text{ L}} = 0.455 M$$

Notes:

Key Words/Concepts

- Molarity (mol/L)
- Solution stoichiometry
    - Volumetric analysis (titration)
        - End point
        - Equivalence point
    - Gravimetric analysis

Notes:

Ch 4/ PowerPoint Study-4 Solution Stoichiometry

Name: _____

Answer these questions as you are watching the videos. They are due in class.
These questions are not just for you to answer but also to prepare you for the exam.
Make sure you understand what you are writing and not just copy from the text book. **Show all work.**

1. What is the concentration of chloride ions in a 0.35 M solution of calcium chloride.
   Strategy:

   a. What is the formula for calcium chloride? _____

   b. How many mols of chloride in one mol of calcium chloride? _____

   c. Set up dimensional analysis and find the molarity of chloride ions.

2. Gravimetric: A 0.4078 g precipitate of lead (II) carbonate forms when a solution of sodium carbonate is added to 100.0 mL solution of the lead (II) solution. What is the concentration of a Pb^{2+} solution? The balanced equation is given below. (Ans: 0.01527 M)

   $$Pb^{2+}(aq) + Na_2CO_3(aq) \rightarrow PbCO_3(s) + 2Na^+(aq)$$

   Strategy:

   a. Convert 0.4078 g of lead (II) carbonate into mols using the formula mass of lead (II) carbonate.

   b. What is the mol ratio of lead (II) ions with lead (II) carbonate? _____

   c. What are the mols of lead ions in the solution? (*Hint: in this case same as in a*) _____

   d. Find the molarity using the mols from c) and the volume (in L) given in the problem.
   (you can also set up the entire answer as dimensional analysis)

3. Neutralization: A 34.62 mL of 0.1510 M NaOH was needed to neutralize 50.0 mL of an H_2SO_4 solution. What is the concentration of the original sulfuric acid solution? (Ans: 0.0523 M)

   a. Write the equation between NaOH and H_2SO_4 and balance it;

   b. Set up the dimensional analysis to find the molarity of sulfuric acid.
   (Strategy: find mol of NaOH → mol ratio to H_2SO_4 → divide by L of H_2SO_4

Chapter 4: Reactions in Aqueous Solutions

4. Zinc reacts with hydrochloric acid to yield hydrogen gas. What mass of hydrogen gas is produced when a 500. mL of 1.200M HCl reacts with 7.35 g of zinc? (Ans: 0.225 g)

$Zn(s) + 2HCl(aq) \rightarrow ZnCl_2(aq) + H_2(g)$

Strategy:

a. Calculate mols of both Zn and HCl and find the mols of H2 formed using mol ratio from the equation. (*Hint: This is a limiting reagent problem*).

b. Use the smaller mols from above and calculate mass of hydrogen produced.

Ch 4/Worksheet/Solution Electrolytes-Ppt Equations   Name: _____

1. Identify the following substances as strong, weak or non electrolytes.
   a. HCl
   b. NH_3
   c. $C_6H_{12}O_6$ (glucose)
   d. N_2

   e. KCl
   f. CaF_2
   g. HNO_3
   h. NaOH

2. Circle the salts that will be soluble in water.
   a. AgBr
   b. $AgNO_3$
   c. K_2SO_4

   d. CuS
   e. $Mg(OH)_2$
   f. $PbSO_4$

3. Indicate whether a precipitate will form when the following solutions are mixed. Write the molecular equation.
   a. iron(III) nitrate and potassium hydroxide

   b. sodium sulfide and nickel(II) sulfate

4. Classify the reaction type, write the products and all the equations in the table below:
   $AgNO_3(aq) + K_2Cr_2O_7(aq) \longrightarrow$

Molecular Equation	
Total Ionic Equation	
Net Ionic Equation	

Chapter 4: Reactions in Aqueous Solutions

5. Write the complete reaction between the following compounds. Write the molecular, ionic and net ionic equation for all the reactions.
   a. Lead (II) nitrate and sodium chloride

   b. ammonium chloride and lithium carbonate

Ch 4/Worksheet/Acid Base Equations

Name: _____

1. Classify each of the following substances as: 1) acid, base, or neutral, and 2) strong or weak.

   a. HNO_3 b. $HClO$ c. NH_3 d. $NaNO_3$ e. $Ba(OH)_2$

2. Complete the following equations. Classify the reaction type, write the products and all the equations.

   a. $Ca(OH)_2(aq) + HCl(aq) \longrightarrow$

Molecular Equation	
Total Ionic Equation	
Net Ionic Equation	

   b. $NaOH(aq) + H_2SO_4(aq) \longrightarrow$

Molecular Equation	
Total Ionic Equation	
Net Ionic Equation	

Ch 4/Worksheet/Redox Equations

Name: _____

1. Write the oxidation numbers of the underlined element in the following compounds.
   a. $\underline{Cr}Cl_3$
   b. $K\underline{Mn}O_4$
   c. $K_2\underline{Cr}O_4$
   d. $\underline{C}H_4$

2. Write the oxidation numbers of the elements underlined in the following equations. Indicate which one is getting oxidized and which one is getting reduced. Also write which is the oxidizing agent in the reactions.

   a. $H\underline{Cl} + \underline{O}_2 \longrightarrow \underline{Cl}_2 + H_2\underline{O}$

   b. $\underline{Ag} + \underline{H}^+ + \underline{NO}_3^- \longrightarrow \underline{Ag}^+ + \underline{H}_2O + \underline{NO}$

   c. $\underline{Se}O_3^{2-} + \underline{I}^- + H^+ \longrightarrow \underline{Se} + \underline{I}_2 + H_2O$

Ch 4/Worksheet/Solution Chemistry   Name: _____

Concentration of Ions

1. What is the concentration of all the ions in a 0.750 M solution of aluminum sulfate?   *(Ans: 3.75 M)*

2. A 4.691 g sample of $MgCl_2$ is dissolved in enough water to give 750. mL of solution. What is the magnesium ion concentration in this solution?   *(Ans: 6.56×10^{-2} M)*

Calculating Molarity:

3. What is the molarity of a solution that has 235 g glucose ($C_6H_{12}O_6$) in 5.00 L water?   *(Ans: 0.261 M)*

4. How many moles of glucose are in 1.89 L of the solution in question 3?   *(Ans: 0.494 mol)*

5. What is the molarity of a solution prepared by dissolving 10.7 g NaI in 0.250 L ?   *(Ans: 0.285 M)*

6. How many grams of KCl are needed to make 50.0 mL of 2.45 M KCl? (Ans: 9.13g)

7. How many grams of LiF would be present in 575 mL of 0.750 M LiF solution? (Ans: 11.2g)

Dilution

8. What volume of 6.0 M H_2SO_4 is needed to make 500.00 mL of 0.25 M H_2SO_4 solution? (Ans: 0.0208L)

9. What molarity should the stock solution be if you want to dilute 25.0 mL to 2.00 L and have the final concentration be 0.103 M ? (Ans: 8.24 M)

10. If you add 4.00 mL of pure water to 6.00 mL of 0.750 M NaCl solution, what is the concentration of sodium chloride in the diluted solution? (Ans: 0.450 M)

Ch 4/Worksheet/Solution Stoichiometry   Name: _____

Gravimetric Analysis/Precipitation Reactions

1. A solution containing water soluble salt of radioactive element thorium can be titrated with oxalic acid solution to form insoluble thorium (IV) oxalate. If 25.00 mL of Thorium (IV) ions requires 19.63 mL of 0.02500 M $H_2C_2O_4$ (Oxalic acid) for complete precipitation (equation given below), then what is the molar concentration of Th^{4+}?   *(Ans: 9.815 x 10^{-3} M)*

$$Th^{4+} + 2H_2C_2O_4 \longrightarrow Th(HC_2O_4)_2 + 2H^+$$

2. A sample of NaCl weighing 1.4477 g is dissolved in 250 mL water, what volume of this solution will be required to titrate 25.00 mL of 0.1000 M $AgNO_3$?   *(Ans: .02526 L)*

Acid – Base Titration

3. How many mL of 0.0195 M hydrochloric acid are required to titrate 25.00 mL of 0.036 M potassium hydroxide?   *(Ans: 46.2 mL)*

4. How many mL of 0.0195 M hydrochloric acid are required to titrate 10.00 mL of 0.0116 M calcium hydroxide?   *(11.9 mL)*

Chapter 4: Reactions in Aqueous Solutions

5. A solution of 0.336 g of KHP, made in water, is titrated and neutralized with 19.67 mL of an unknown concentration of sodium hydroxide solution. What is the molarity of a sodium hydroxide? (Molecular Weight of KHP = 204.23 g/mol and mol ratio of KHP and sodium hydroxide is 1:1). *(Ans: 0.0836 M)*

6. How many grams of potassium hydrogen phthalate (KHP) are needed to neutralize 22.36 mL of a 0.1205 M solution of sodium hydroxide? (Molecular Weight of KHP = 204.23 g/mol and mol ratio of KHP and sodium hydroxide is 1:1). *(Ans: 0.550g)*

7. In a reaction like neutralization of stomach acid, 175 mL of 1.55 M sodium hydrogen carbonate is added to 235 mL of 1.22 M hydrochloric acid. A) How much carbon dioxide is liberated? B) What is the molarity of sodium chloride produced? (assume that the volumes are additive) *(Ans: 11.9g CO_2, 0.662 M)*

Chapter 5
Gases

Properties of Gases:
1. Gases fill their containers completely (unlike liquids).
2. Gases occupy far more space than liquids or solids because their molecules have a large amount of space between them.
3. Gases are compressible.

Kinetic Molecular Theory:
1. All molecules are in constant motion.
2. Molecules keep traveling in straight line until they collide with another and change course (this is the reason molecules are considered elastic).
3. Molecules collide with other molecules in an elastic collision so there is no net gain or loss of energy.
4. There is always some empty space between particles so no container is completely packed with gas molecules.
5. The kinetic energy of the molecules depend on the temperature; so warm gas molecules move faster than colder gases.

Pressure:
Force/unit area , units: Pascal, atmosphere, torr, mmHg, newtons (SI unit)
Measuring atmospheric pressure: barometer, using mercury.
760 mmHg = 1 atm = 760 torr = 101.325 kPa = 14.696 psi
manometer: measuring gas pressures.

Gas Laws:

Boyle's Law	P inversely proportional to V	PV = constant	$P_1V_1 = P_2V_2$
Charles's Law	V directly proportional to T	V = T • constant	$V_1/T_1 = V_2/T_2$
Avogradro's Law	V directly proportional to n at constant P and T	V = n • constant	$V_1/n_1 = V_2/n_2$
Dalton's Law	Total pressure of mixture of gases is equal to the partial pressures of all the gases in the system.		$P_{total} = P_1 + P_2 + P_3 + \ldots$

Temperature is always measured in Kelvin (0°C = 273.15 K)
STP conditions: T = 273.15 K and P = 1 atm, n = 1 mol, V = 22.4 L

Combined Gas Law	$P_1V_1/n_1T_1 = P_2V_2/n_2T_2$
Ideal Gas Law	PV = nRT (R = gas constant = 0.082058 L atm/mol K) (Values of R changes with different units)
Gas Density	d = MP/RT (M = molar mass)

Key Concepts and Words:

Manometer	Kinetic Molecular Theory	The Laws: Boyle's, Charles's, Avogadro's, Combined Gas Law, Ideal Gas Law, Gay-Lussac's, Dalton's.
Molar Gas Constant	Standard Pressure, Temperature and Volume	Diffusion and Effusion
Real Gas	Gas Stoichiometry	Collecting gas over water

Chapter 5 Gases -1 Gas Properties and Pressure

Dr. Sapna Gupta

Properties of Gases

1. They are compressible.
2. They expand to fill the container.
3. Pressure, volume, temperature, and amount are related.
4. They have low density.
5. Form a homogeneous mixture.

Notes:

Kinetic Molecular Theory

A theory, developed by physicists, that is based on the assumption that a gas consists of molecules in constant random motion.

Postulates of the Kinetic Theory

1. Gases are composed of molecules whose sizes are negligible.
2. Molecules move randomly in straight lines in all directions and at various speeds.
3. The forces of attraction or repulsion between two molecules (intermolecular forces) in a gas are very weak or negligible, except when the molecules collide.
4. When molecules collide with each other, the collisions are elastic.
5. The average kinetic energy of a molecule is proportional to the absolute temperature.

Notes:

Pressure

Pressure is force exerted per unit area. P = Force/Area

The SI unit for pressure is the pascal, Pa.
Other Units
  atmosphere, atm
  mmHg
  torr
  bar

$1 \text{ Pa} = 1 \text{ N/m}^2$ →

1 atm*
101,325 Pa
760 mmHg*
760 torr*
1.01325 bar
14.7 psi
*These are exact numbers.

Notes:

Measurement of Pressure

A **barometer** is a device for measuring the pressure of the atmosphere.

A **manometer** is a device for measuring the pressure of a gas or liquid in a vessel.

Notes:

Chapter 5: Gases

Key Points

- Properties of gases
- The kinetic molecular theory
- Gas pressure
  - Units
  - Calculation
  - Measurement

Notes:

Ch 5/ PowerPoint Study-1 Gases-Units and KMT

Name: _____

Answer these questions as you are watching the videos. They are due in class.
These questions are not just for you to answer but also to prepare you for the exam.
Make sure you understand what you are writing and not just copy from the text book. **Show all work.**

1. Describe pressure? Give three different units used in chemistry for pressure with one of them being the SI unit (circle it).

2. Carry out the following conversions of the pressure unit (show all work for credit)

a) 747mmHg into atm	b) 0.245 atm into torr	c) 1.067×10^6 Pa into atm

3. Give the five postulates of the kinetic theory in your own words.

Chapter 5   Gases -2 Gas Laws

Dr. Sapna Gupta

Gas Laws

- Gas laws – empirical relationships among gas parameters.
  - Volume (V)
  - Pressure (P)
  - Temperature (T)
  - Amount of a gas (n)

- By holding two of these physical properties constant, it becomes possible to show a simple relationship between the other two properties.

Boyle's Law

- The volume of a sample of gas is inversely proportional to pressure at constant temperature. (one increases if the other decreases and *vice versa*)

$$V \propto \frac{1}{P}$$

A sample of a gas at 0°C is placed in 2.0 L container at a pressure of 1.0 atm.

1.0 atm
2.0 L

When the volume of the gas is reduced to 1.0 L the pressure on the gas is doubled to 2.0 atm.

2.0 atm
1.0 L

$$PV = \text{constant}$$
$$P_1V_1 = P_2V_2$$

Boyle's Law Contd....

Graphical representation

Source: Sapna Gupta   Source: Sapna Gupta

Solved Problem:
A volume of oxygen gas occupies 38.7 mL at 751 mmHg and 21°C. What is the volume if the pressure changes to 359 mmHg while the temperature remains constant?

$V_i = 38.7$ mL $\qquad V_f = ?$
$P_i = 751$ mmHg $\qquad P_f = 359$ mmHg
$T_i = 21°$ C $\qquad T_f = 21°$ C

$$V_f = \frac{P_i V_i}{P_f}$$

$$V_f = \frac{(38.7 \text{ mL})(751 \text{ mmHg})}{(359 \text{ mmHg})}$$

= 81.0 mL
(3 significant figures)

Notes:

Notes:

Chapter 5: Gases — 167

Charles's Law

- The volume of a sample of gas at constant pressure is directly proportional to the absolute temperature (K).

$$V \propto T$$

$$\frac{V}{T} = \text{constant}$$

$$\frac{V_i}{T_i} = \frac{V_f}{T_f}$$

Notes:

Charles's Law contd....

Graphical Representation

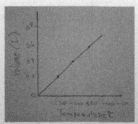

Source: Sapna Gupta

The temperature -273.15°C is called **absolute zero**. It is the temperature at which the volume of a gas is hypothetically zero.

This is the basis of the absolute temperature scale, the Kelvin scale (K).

Notes:

Chapter 5: Gases

Solved Problem:
You prepared carbon dioxide by adding HCl(*aq*) to marble chips, $CaCO_3$. According to your calculations, you should obtain 79.4 mL of CO_2 at 0°C and 760 mmHg. How many milliliters of gas would you obtain at 27°C?

V_i = 79.4 mL V_f = ?
P_i = 760 mmHg P_f = 760 mmHg
T_i = 0° C = 273 K T_f = 27° C = 300. K

$$V_f = \frac{T_f V_i}{T_i}$$

$$V_f = \frac{(300.\ K)(79.4\ mL)}{(273\ K)}$$

= 87.3 mL
(3 significant figures)

Notes:

Avogadro's Law

The volume of a gas sample is directly proportional to the number of moles in the sample at constant pressure and temperature.

$V \propto n$

$\frac{V}{n} = constant$

$$\frac{V_i}{n_i} = \frac{V_f}{n_f}$$

$3H_2(g)$ + $N_2(g)$ ⟶ $2NH_3(g)$

3 molecules	+	1 molecule	⟶	2 molecules
3 moles	+	1 mole	⟶	2 moles
3 volumes	+	1 volume	⟶	2 volumes

Notes:

Dalton's Law of Partial Pressure

- Dalton found that in a mixture of unreactive gases, each gas acts as if it were the only gas in the mixture as far as pressure is concerned.
- The sum of the partial pressures of all the different gases in a mixture is equal to the total pressure of the mixture:

$$P = P_A + P_B + P_C + \ldots$$

Partial Pressure

The pressure exerted by a particular gas in a mixture, e.g. in a mixture of nitrogen and oxygen will be addition of the pressure of the two gases.

$$P_{total} = P_{N_2} + O_2$$

Notes:

Solved Problem:
A 100.0-mL sample of air exhaled from the lungs is analyzed and found to contain 0.0830 g N_2, 0.0194 g O_2, 0.00640 g CO_2, and 0.00441 g water vapor at 35°C. What is the partial pressure of each component and the total pressure of the sample?

$$P_{N_2} = \frac{\left(0.0830 \text{ g } N_2 \cdot \frac{1 \text{ mol } N_2}{28.01 \text{ g } N_2}\right)\left(0.08206 \frac{L \cdot atm}{mol \cdot K}\right)(308 \text{ K})}{(100.0 \text{ mL})\left(\frac{1 L}{10^{-3} \text{ mL}}\right)} = 0.749 \text{ atm}$$

$$P_{O_2} = \frac{\left(0.0194 \text{ g } O_2 \cdot \frac{1 \text{ mol } O_2}{32.00 \text{ g } O_2}\right)\left(0.08206 \frac{L \cdot atm}{mol \cdot K}\right)(308 \text{ K})}{(100.0 \text{ mL})\left(\frac{1 L}{10^{-3} \text{ mL}}\right)} = 0.153 \text{ atm}$$

$$P_{CO_2} = \frac{\left(0.00640 \text{ g } CO_2 \cdot \frac{1 \text{ mol } CO_2}{44.01 \text{ g } CO_2}\right)\left(0.08206 \frac{L \cdot atm}{mol \cdot K}\right)(308 \text{ K})}{(100.0 \text{ mL})\left(\frac{1 L}{10^{-3} \text{ mL}}\right)} = 0.0368 \text{ atm}$$

$$P_{H_2O} = \frac{\left(0.00441 \text{ g } H_2O \cdot \frac{1 \text{ mol } H_2O}{18.01 \text{ g } H_2O}\right)\left(0.08206 \frac{L \cdot atm}{mol \cdot K}\right)(308 \text{ K})}{(100.0 \text{ mL})\left(\frac{1 L}{10^{-3} \text{ mL}}\right)} = 0.0619 \text{ atm}$$

$$P = P_{N_2} + P_{O_2} + P_{CO_2} + P_{H_2O} = 1.00 \text{ atm}$$

Notes:

Chapter 5: Gases

Key Points

- The gas laws
  - Boyle's law
  - Charles' law
  - Avogadro's law
  - Dalton's law

Notes:

Ch 5/ PowerPoint Study-2 Gases-Gas Laws   Name: _____

Answer these questions as you are watching the videos. They are due in class.
These questions are not just for you to answer but also to prepare you for the exam.
<u>*Make sure you understand what you are writing and not just copy from the text book.*</u> **Show all work.**

1. If 25.5 L of oxygen are cooled from 150°C to 50°C at constant pressure, what is the new volume of oxygen?
   (Ans: 19.5 L)

2. A flexible vessel contains 47.0 L of gas where the pressure is 1.30 atm. What will the volume be when the pressure is 0.850 atm, the temperature remaining constant?   (Ans: 72 L)

3. A gas occupying a volume of 1.50 L exerts a pressure of 700 mmHg at 200°C. What is the correct pressure at 6.00 L and 200°C?   (175 mmHg)

4. A rigid container is charged with a gas to a pressure of 760 mmHg at 20.0°C and tightly sealed. If the temperature of the gas *increases* by 40.0°C what is the new pressure?   (Ans: 864 mmHg)

5. A mixture of gases contains nitrogen at a partial pressure of 0.50 atmospheres, oxygen at a partial pressure of 0.20 atmospheres and carbon dioxide. The total gas pressure is 0.80 atmospheres.
   a. Find the partial pressure of carbon dioxide.   (Ans: 0.10 atm)
   b. If there are 0.25 moles of nitrogen, what is the number of moles of oxygen?   (Ans: 0.10 mol)

Chapter 5 Gases - 2 Combined Gas Law, Ideal Gas Law and Applications of Gas Laws
Dr. Sapna Gupta

Combined Gas Law

The volume of a sample of gas at constant pressure is inversely proportional to the pressure and directly proportional to the absolute temperature.

The mathematical relationship: $V \propto \dfrac{T}{P}$

In equation form: $\dfrac{PV}{T} = \text{constant}$

$$\dfrac{P_i V_i}{T_i} = \dfrac{P_f V_f}{T_f}$$

Include the Avogadro's law and it becomes (1=initial and 2=final)

$$\boxed{\dfrac{P_i V_i}{n_i T_i} = \dfrac{P_f V_f}{n_f T_f}} \quad \text{OR} \quad \boxed{\dfrac{P_1 V_1}{n_1 T_1} = \dfrac{P_2 V_2}{n_2 T_2}}$$

Solved Problem:
Divers working from a North Sea drilling platform experience pressure of 5.0×10^1 atm at a depth of 5.0×10^2 m. If a balloon is inflated to a volume of 5.0 L (the volume of the lung) at that depth at a water temperature of 4°C, what would the volume of the balloon be on the surface (1.0 atm pressure) at a temperature of 11°C?

$V_i = 5.0$ L $V_f = ?$
$P_i = 5.0 \times 10^1$ atm $P_f = 1.0$ atm $V_f = \dfrac{T_f P_i V_i}{T_i P_f}$
$T_i = 4°\text{C} = 277$ K $T_f = 11°\text{C} = 284$ K

$$V_f = \dfrac{(284\,\text{K})(5.0 \times 10^1\,\text{atm})(5.0\,\text{L})}{(277\,\text{K})(1.0\,\text{atm})} = 2.6 \times 10^2\,\text{L}$$

(2 significant figures)

Notes:

Ideal Gas Law

Ideal Gas Law
- Combining the historic gas laws yields:

    Boyle's law: $V \propto \dfrac{1}{P}$

    Charles's law: $V \propto T$ $\longrightarrow$ $V \propto \dfrac{nT}{P}$

    Avogadro's law: $V \propto n$

- Adding the proportionality constant, R

$$V = R\dfrac{nT}{P} \longrightarrow \boxed{PV = nRT}$$

$$R = \dfrac{PV}{nT}$$

Values of R
0.08206 L atm/(mol K)
8.3145 J/(mol K)
8.3145 kg m²/(s² mol K)
1.987 cal/mol K

Notes:

Standard Temperature and Pressure (STP)

- The reference condition for gases, chosen by convention to be exactly **0°C and 1 atm pressure**.
- The ideal gas equation is *not* exact, but for most gases it is quite accurate near STP (760 torr (1 atm) and 273 K)
- An "ideal gas" is one that "obeys" the ideal gas equation.
- At STP, **1 mol** of an ideal gas occupies **22.41 L**.

    at STP
    T = 0°C = 273 K
    n = 1 mol
    P = 1 atm
    V = 22.41 L

Notes:

Chapter 5: Gases

Solved Problem:
A steel cylinder with a volume of 68.0 L contains O_2 at a pressure of 15,900 kPa at 23ºC. What is the volume of this gas at STP?

$P_1 = 15,900 \text{ kPa} \times \dfrac{1 \text{ atm}}{101.3 \text{ kPa}} = 157.0 \text{ atm}$

$T_1 = 23 + 273 = 296 \text{ K}$
$V_1 = 68.0 \text{ L}$

$P_2 = 1 \text{ atm}$
$T_2 = 273 \text{ K}$
$V_2 = ?$

$$\dfrac{P_1 V_1}{n_1 T_1} = \dfrac{P_2 V_2}{n_2 T_2}$$

$$V_2 = \dfrac{P_1 V_1 T_2}{T_1 P_2} = \dfrac{(157.0 \text{ atm})(68.0 \text{ L})(273 \text{ K})}{(296 \text{ K})(1 \text{ atm})} = 9850 \text{ L}$$

Notes:

Solved Problem:
A 50.0-L cylinder of nitrogen, N_2, has a pressure of 17.1 atm at 23°C. What is the mass of nitrogen in the cylinder?

$V = 50.0 \text{ L}$
$P = 17.1 \text{ atm}$
$T = 23°C = 296 \text{ K}$

$n = \dfrac{PV}{RT}$

$$n = \dfrac{(17.1 \text{ atm})(50.0 \text{ L})}{\left(0.08206 \dfrac{L \bullet atm}{mol \bullet K}\right)(296 \text{ K})} = 35.20 \text{ mols}$$

$\text{mass} = 35.20 \text{ mol} \dfrac{28.02 \text{ g}}{\text{mol}}$

mass = 986 g
   (3 significant figures)

Notes:

Solved Problem:
For an ideal gas, calculate the pressure of the gas if 0.215 mol occupies 338 mL at 32.0°C.

$n = 0.215$ mol
$V = 338$ mL $= 0.338$ L
$T = 32.0 + 273.15 = 305.15$ K
$P = ?$

$$PV = nRT \Rightarrow P = \frac{nRT}{V}$$

$$P = \frac{(0.215 \text{ mol})\left(0.08206 \frac{\text{L} \times \text{atm}}{\text{mol} \times \text{K}}\right)(305.15 \text{ K})}{0.338 \text{ L}} = 15.928 = 15.9 \text{ atm}$$

Gas Density and Molar Mass

- Using the ideal gas law, it is possible to calculate the moles in 1 L at a given temperature and pressure. The number of moles can then be converted to grams (per liter).

Relation to density
- Get n/V on one side (mol/vol) $\quad \frac{n}{V} = \frac{P}{RT}$

- Multiply by molar mass M (g/mol) $\quad \boxed{\mathcal{M} \times \frac{n}{V} = \frac{P}{RT} \times \mathcal{M}}$

- Units of density are g/L here. $\quad d = \frac{P\mathcal{M}}{RT}$

To find molar mass, rearrange the above equation.

$$\mathcal{M} = \frac{dRT}{P}$$

Chapter 5: Gases

Solved Problem:
What is the density of methane gas (natural gas), CH_4, at 125°C and 3.50 atm?

$M_m = 16.04$ g/mol
$P = 3.50$ atm
$T = 125°$ C = 398 K

$$d = \frac{M_m P}{RT}$$

$$d = \frac{(16.04 \frac{g}{mol})(3.50 \text{ atm})}{\left(0.08206 \frac{L \cdot atm}{mol \cdot K}\right)(398 \text{ K})}$$

$$d = 1.72 \frac{g}{L}$$

(3 significant figures)

Notes:

All Gas Laws

- Gases are compressible because the gas molecules are separated by large distances.

- The magnitude of P depends on how often and with what force the molecules strike the container walls.

- At constant T, as V increases, each particle strikes the walls less frequently and P decreases. (Boyle's Law)

- To maintain constant P, as V increases T must increase; fewer collisions require harder collisions. (Charles' Law)

- To maintain constant P and T, as V increases n must increase. (Avogadros' Law)

- Gas molecules do not attract or repel one another, so one gas is unaffected by the other and the total pressure is a simple sum. (Dalton's Law)

Notes:

Chapter 5: Gases

Key Points

- Combined gas law
- The ideal gas law
- Standard Temperature and Pressure
- Applications of gas laws
  - Density
  - Molar mass

Notes:

Ch 5/ PowerPoint Study-3Gases-Combined Gas Law/   Name: _____

Ideal Gas Law/Applications of Gas Laws
Answer these questions as you are watching the videos. They are due in class.
These questions are not just for you to answer but also to prepare you for the exam.
Make sure you understand what you are writing and not just copy from the text book. **Show all work.**

Combined Gas Law and Ideal Gas Law

1. A 24.0 liter sample of pure nitrogen gas at 20.0°C and 1.50 atmospheres pressure is heated. What is its pressure at 313°C if its volume is 36.0 liters? (Ans: 2.00 atm)
   (Think: do you need to convert centigrade to Kelvin? Ans: yes!)

2. Find the number of grams of pentane (C_5H_{12}) gas in a 11.2 L sample at 0°C and 2.40 atm pressure. (Ans: 86.3 g)
   (Think: do you need to convert centigrade to Kelvin?)

3. A 3.80-L cylinder contains 6.83 g of methane, CH_4, at a pressure of 3320 mmHg. What is the temperature of the gas? (Ans: 201 °C)
   (Hint: calculate the moles of methane and then use the ideal gas law)

4. Calculate the density of carbon dioxide at 546 K and 4.00 atmospheres pressure. (Ans: 3.93 g/L)

Chapter 5   Gases - 4 Gas Stoichiometry

Dr. Sapna Gupta

Stoichiometry in Gases

Amounts of gaseous reactants and products can be calculated by utilizing
- The ideal gas law to relate moles to T, P and V.
- Moles can be related to mass by the molar mass
- The coefficients in the balanced equation to relate moles of reactants and products

---

Solved Problem:
When a 2.0-L bottle of concentrated HCl was spilled, 1.2 kg of $CaCO_3$ was required to neutralize the spill. What volume of CO_2 was released by the neutralization at 735 mmHg and 20.°C?

First, write the balanced chemical equation:

$CaCO_3(s) + 2HCl(aq) \rightarrow CaCl_2(aq) + H_2O(l) + CO_2(g)$

Second, calculate the moles of CO_2 produced:
Molar mass of $CaCO_3$ = 100.09 g/mol

$$1.2 \times 10^3 \text{ g CaCO}_3 \cdot \frac{1 \text{ mol CaCO}_3}{100.09 \text{ g CaCO}_3} \cdot \frac{1 \text{ mol CO}_2}{1 \text{ mol CaCO}_3} = 11.99 \text{ mol}$$

n = 11.99 mol
P = 735 mmHg
 = 0.967 atm
T = 20° C = 293 K

$$V = \frac{nRT}{P}$$

$$V = \frac{(11.99 \text{ mol})\left(0.08206 \frac{L \cdot atm}{mol \cdot K}\right)(293 \text{ K})}{(0.967 \text{ atm})} = 2.98 \times 10^2 \text{ L}$$

(3 significant figures)

Collecting Gas Over Water

- Gases are often collected over water. The result is a mixture of the gas and water vapor.
- The total pressure is equal to the sum of the gas pressure and the vapor pressure of water.
- The partial pressure of water depends only on temperature and is known (Table 5.6).
- The pressure of the gas can then be found using Dalton's law of partial pressures.

Temperature (°C)	Pressure (mmHg)
0	4.6
10	9.2
15	12.8
17	14.5
19	16.5
21	18.7
23	21.1
25	23.8
27	26.7
30	31.8
40	55.3
60	149.4
80	355.1

$P = P_{H_2} + P_{H_2O}$
$P_{H_2} = P - P_{H_2O}$
$P_{H_2} = 769 \text{ mmHg} - 16.5 \text{ mmHg}$

$P_{H_2} = 752.5 \text{ mmHg}$

$P_{H_2} = 753 \text{ mmHg}$
(no decimal places)

Chapter 5: Gases

Solved Problem:
You prepare nitrogen gas by heating ammonium nitrite:
$$NH_4NO_2(s) \rightarrow N_2(g) + 2H_2O(l)$$
If you collected the nitrogen over water at 23°C and 727 mmHg, how many liters of gas would you obtain from 5.68 g NH_4NO_2?

P = 727 mmHg
P_{vapor} = 21.1 mmHg   Molar mass NH_4NO_2 = 64.06 g/mol
P_{gas} = 706 mmHg
T = 23° C = 296 K

$$5.68 \text{ g } NH_4NO_2 \cdot \frac{1 \text{ mol } NH_4NO_2}{64.04 \text{ g } NH_4NO_2} \cdot \frac{1 \text{ mol } N_2}{1 \text{ mol } NH_4NO_2} = 0.08887 \text{ mol } N_2$$

n = 0.0887 mol

$$V = \frac{nRT}{P} \qquad V = \frac{(0.0887 \text{ mol})\left(0.08206 \frac{L \cdot atm}{mol \cdot K}\right)(296 \text{ K})}{\left(706 \text{ mmHg} \cdot \frac{1 \text{ atm}}{760 \text{ mmHg}}\right)} = 2.32 \text{ L of } N_2$$
(3 significant figures)

Notes:

Solved Problem:
Oxygen was produced and collected over water at 22°C and a pressure of 754 torr.
$$2 \text{ KClO}_3(s) \rightarrow 2 \text{ KCl}(s) + 3 \text{ O}_2(g)$$
325 mL of gas were collected and the vapor pressure of water at 22°C is 21 torr. Calculate the number of moles of O_2 and the mass of $KClO_3$ decomposed.

$P_{total} = P_{O_2} + P_{H_2O} = P_{O_2} + 21 \text{ torr} = 754 \text{ torr}$

$P_{O_2} = 754 \text{ torr} - 21 \text{ torr} = 733 \text{ torr} = 733/760 \text{ atm}$

$V = 325 \text{ mL} = 0.325 \text{ L}$

$T = 22°C + 273 = 295 \text{ K}$

$$n_{O_2} = \frac{(733/760 \text{ atm})(0.325 \text{ L})}{\left(0.08206 \frac{L \cdot atm}{mol \cdot K}\right)(295 \text{ K})} = 1.29 \times 10^{-2} \text{ mol } O_2$$

$$2 \text{ KClO}_3(s) \rightarrow 2 \text{ KCl}(s) + 3 \text{ O}_2(g)$$

$$1.29 \times 10^{-2} \text{ mol } O_2 \left(\frac{2 \text{ mol KClO}_3}{3 \text{ mol } O_2}\right)\left(\frac{122.6 \text{ g KClO}_3}{1 \text{ mol KClO}_3}\right) = 1.06 \text{ g KClO}_3$$

Notes:

Speed of Gas

- Root mean square (rms) speed (u_{rms}) $\quad u_{rms} = \sqrt{\dfrac{3RT}{\mathcal{M}}}$

- For two gases (1 and 2) $\quad \dfrac{u_{rms}(1)}{u_{rms}(2)} = \sqrt{\dfrac{\mathcal{M}_2}{\mathcal{M}_1}}$

- Effect of Temperature on Molecular Speed (1st graph)
- Effect of Molar Mass on Molecular Speed (2nd graph)

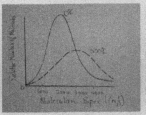

Source: Sapna Gupta

Source: Sapna Gupta

Diffusion and Effusion

Diffusion	Effusion
The process whereby a gas spreads out through another gas to occupy the space uniformly. Below molecules diffuses through air.	The process by which a gas flows through a small hole in a container. A pinprick in a container is one example of effusion.

Source: Sapna Gupta

Source: Sapna Gupta

Real Gases

At high pressure the relationship between pressure and volume does not follow Boyle's law. This is illustrated on the graph below.

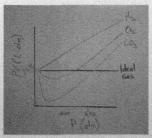

Source: Sapna Gupta

At high pressure, some assumptions of the kinetic theory no longer hold true. At high pressure:
1. the volume of the gas molecule is not negligible.
2. the intermolecular forces are not negligible.

Van der Waal's Equation

An equation that is similar to the ideal gas law, but which includes two constants, *a* and *b*, to account for deviations from ideal behavior.

The term V becomes $(V - nb)$ to account for the space between molecules.

The term P becomes $(P + n^2a/V^2)$ to account for attraction/repulsion between molecules.

Values for *a* and *b* are different for different gases and can be found in Table 5.7

- The ideal gas law $PV = nRT$

  becomes van der Waal's equation
  $$\left(P + \frac{an^2}{V^2}\right)(V - nb) = nRT$$
  a and *b* have specific values for each gas

  (corrected pressure term) (corrected volume term)

Chapter 5: Gases

Key Points

- Gas stoichiometry
- Collecting gas over water
- Gas mixtures
  - Molecular speed
  - Diffusion and effusion
- Deviation from ideal behavior
  - Factors causing deviation
  - Van der Waal's equation

Notes:

Ch 5/ PowerPoint Study-4Gases-Gas Stoichiometry   Name: _____

Answer these questions as you are watching the videos. They are due in class.
These questions are not just for you to answer but also to prepare you for the exam.
Make sure you understand what you are writing and not just copy from the text book. **Show all work.**

1. What volume of ammonia gas, measured at 660.3 mmHg and 58.2°C, is required to produce 6.46 g of ammonium sulfate according to the following balanced chemical equation?   (Ans: 3.07 L)

$$2NH_3(g) + H_2SO_4(aq) \longrightarrow (NH_4)_2SO_4(s)$$

Strategy:
a. Calculate the mols of ammonium sulfate.

b. What is the mol ratio of ammonium sulfate to ammonia?

c. Use the mols from (b) to substitute in the ideal gas law to find volume of ammonia gas.

2. What grams of $CaCO_3$ will produce 34.8 L volume of CO_2 gas at 645 torr and 800 K according to the equation $CaCO_3(s) \rightarrow CaO(s) + CO_2(g)$?   (Ans: 44.9 g)

Strategy:
a. Use the ideal gas law to find the mols of carbon dioxide.

b. Find the mol ratio of carbon dioxide to calcium carbonate.

c. Find the grams in mols of calcium carbonate from (b).

(We will do "gases over water" in class practice)

Ch 5/Worksheet/Gases-Gas Laws

Name: _____

Introduction

1. Which is not a property of a gas?
   a. Density varies with temperature
   b. Assumes the shape an volume of its container
   c. Are compressible
   d. Density is larger than that of a liquid
   e. Form homogeneous mixtures with one another

2. Hydrogen gas exerts a pressure of 466 torr in a container. What is this pressure in atmospheres (1 atm = 101,325 Pa = 760 torr)? (Ans: 0.613 atm)

Gas Laws

3. A sample of a gas occupies 1.40×10^3 mL at 25°C and 760 mmHg. What volume will it occupy at the same temperature and 380 mmHg? (Ans: 2800 mL)

4. A sample of a gas has an initial pressure of 0.987atm and a volume of 12.8 L. What is the final pressure if the volume is increased to 25.6 L? (Ans: 0.494 atm)

5. A sample of nitrogen gas has a volume of 32.4 L at 20.0°C. The gas is heated to 220°C at constant pressure. What is the final volume of nitrogen? (Ans: 54.5 L)

Chapter 5: Gases

Combined and Ideal Gas Law

6. If 600. cm^3 of H_2 at 25°C and 750. mm Hg is compressed to a volume of 480. cm^3 at 41°C, what is the new pressure? (Ans: 988 mmHg)

7. At a particular temperature and pressure, 15.0 g of CO_2 occupy 7.16 liters. What is the volume of 12.0 g of CH_4 at the same temperature and pressure? (Ans: 15.7 L)

8. How many liters of methane are there in 8.00 grams at STP? (Ans: 11.2 L)

Application of Gas Laws

9. Calculate the density of chlorine gas at STP. (Ans: 3.17 g/L)

10. What is the molar mass of a gas if 7.00 grams occupy 6.20 liters at 29°C and 760. mm Hg pressure? (Ans: 28.0 g/mol)

Ch 5/Worksheet/Gases—Partial Pressure and Stoichiometry

Name: _____

1. A sample of air collected at STP contains 0.039 moles of N_2, 0.010 moles of O_2, and 0.001 moles of Ar. (Assume no other gases are present.)
   a. Find the partial pressure of O_2.
   b. What is the volume of the container?

2. A sample of hydrogen gas (H_2) is collected over water at 19°C.
   a. What are the partial pressures of H_2 and water vapor if the total pressure is 756 mm Hg?
   b. What is the partial pressure of hydrogen gas in atmospheres?

3. What volume of O_2 at 710. mm Hg pressure and 36.0°C is required to react with 6.52 g of CuS? *(Ans: 3.70 L)*

$$CuS(s) + 2\,O_2(g) \rightarrow CuSO_4(s)$$

Chapter 5: Gases

4. Write a balanced equation for the production of ammonia gas (NH_3) from nitrogen gas (N_2) and hydrogen gas (H_2).

   a. What volume of ammonia is produced from 4.50 liters of H_2 at STP? *(Ans: 3.00 L)*

   b. What mass of ammonia is produced from 5.60 liters of N_2 at STP? *(Ans: 8.50 g)*

   c. What volume of ammonia is produced from 12.1 grams of H_2 at 25°C and 1.00 atmosphere pressure? *(Ans: 98.0 L)*

5. Use the following chemical equation to answer the questions below:

   $$Na(s) + O_2(g) \rightarrow Na_2O(s)$$

   a. How many grams of sodium are needed to completely react with 2.80 liters of O_2 at STP? *(Ans: 11.5 g)*

   b. What volume of O_2 at 25°C and 2.00 atm is needed to completely react with 4.60 grams of sodium? *(Ans: 0.610 L)*

6. To prepare a sample of hydrogen gas, a student reacts 7.78 grams of zinc with acid:

$$Zn(s) + 2\ H^+\ (aq) \rightarrow Zn^{2+}(aq) + H_2(g)$$

The hydrogen is collected over water at 22°C and the total pressure of gas collected is 750. mm Hg. What is the partial pressure of H_2? What volume of wet hydrogen gas is collected? *(Ans: 3.00 L)*

7. Hydrogen peroxide was catalytically decomposed and 75.3 mL of oxygen gas was collected over water at 25°C and 742 torr. What mass of oxygen was collected?
   a. 0.00291 g   b. 0.0931 g   c. 0.0962 g   d. 0.0993 g   e. 0.962 g

Vapor Pressure of Water										
Temp (°C)	15	16	17	18	19	20	21	22	23	24	25
P_{H2O} (mm Hg)	13	14	15	15	16	18	19	20	21	22	24

Chapter 7
Quantum Theory

Quantum View

<u>Electromagnetic Spectrum</u>: based on wavelength and frequency.
<u>Continuous spectrum</u>: white light is continuous spectrum, even when passing through a prism it is continuous as one each color merges into the other.
<u>Line spectrum</u>: obtained by exciting a gas, passing the energy released through a slit and prism. Only certain lines are observed. Also called emission spectrum. Each element has a unique line spectrum hence also known as fingerprint of element.

1900	Max Planck	Derived relationship between energy and frequency. All matter absorbs or emits electromagnetic energy only in discrete quantity called quanta. $E = h\nu$ (energy = Planck's constant x frequency)
1905	Albert Einstein	Photoelectric effect. He called the bundles of energy Photons.
1913	Niels Bohr	Proposed atom has energy levels. Electrons stay in these energy levels and when energy is absorbed there is transition to another level thus energy is given off when electron comes back to the original level. Thus energy is always given off in certain quantities. $E_n = -B/n^2$
1923	Louis de Broglie	Proposed that particles have a wave nature. (led to development of electron microscope) $\lambda = h/mv$
1926	Erwin Schrödinger	Developed the wave function equation which determines the position of an electron in the atom.
1927	Werner Heisenberg	Said that one cannot determine the location of the electron.

<u>Calculations</u>
E = energy   h = Planck's constant ν = frequency   B = constant
n = shell number   m = mass   v = velocity λ = wavelength

Planck's equation $E = h\nu$
Bohr's equation for calculation of energy of electron in an energy level, energy transition and line spectrum:
$E_n = -B/n^2$
De Broglie's equation for calculating wavelength of electron: $\lambda = h/mv$

Concepts

1. Ground state and excited state
2. Quantum numbers
   a. Principal quantum number (n) – shell/orbit number (1, 2, 3, 4 etc)
   b. Orbital angular momentum quantum number (l) – orbital location (s, p, d, f)
   c. Magnetic quantum number (m_l) – shape of orbitals (1 for s, 3 for p, 5 for d and 7 for f)
   d. Electron spin quantum number (m_s) – electron spin

Table 1

Quantum Number	Allowed Values	Name and Meaning
n	$n = 1, 2, 3, \ldots\ldots$	*Principal quantum number*: orbital energy and size.
l	$l = (n-1), (n-2), \ldots, 0$	*Azimuthal (or orbital) quantum number*: orbital shape (and energy in a multi-electron atom), letter name for subshell (s, p, d, f)
m_l	$m_l = l, (l-1), \ldots, 0, \ldots, (-l+1), -l$	*Magnetic quantum number*: orbital orientation
m_s	$m_s = 1/2, -1/2$	*Electron spin quantum number*: spin up (↑) or spin down (↓).

Table 2

l Value	Letter Equivalent to l Value	No. of Orbitals in Set	Approximate Shape of Orbitals with Specific l Values
0	s	1	spherical
1	p	3	p_x, p_y, p_z are dumbbells along x, y, and z axes
2	d	5	mostly cloverleaf shapes
3	f	7	very complicated shapes!

Table 3

Shell (n)	Subshell (l)	Orbital Name (nl)	Orientations (m_l)	No. of Orbitals	Maximum Occupancy
$n = 1$	$l = 0$	1s	$m_l = 0$	1	2 e⁻
$n = 2$	$l = 0$	2s	$m_l = 0$	1	2 e⁻
	$l = 1$	2p	$m_l = 1, 0, -1$ (or p_x, p_y, p_z)	3	6 e⁻
$n = 3$	$l = 0$	3s	$m_l = 0$	1	2 e⁻
	$l = 1$	3p	$m_l = 1, 0, -1$ (or p_x, p_y, p_z)	3	6 e⁻
	$l = 2$	3d	$m_l = 2, 1, 0, -1, -2$ (or $d_{xy}, d_{yz}, d_{xz}, d_{x^2-y^2}, d_{z^2}$)	5	10 e⁻

Chapter 7 Atomic Structure -1 Quantum Model of Atom

Dr. Sapna Gupta

The Electromagnetic Spectrum

- The electromagnetic spectrum includes many different types of radiation which travel in waves.
- Visible light accounts for only a small part of the spectrum
- Other familiar forms include: radio waves, microwaves, X rays

Wave Nature

- **Wavelength**: λ (lambda) distance between identical points on successive waves…peaks or troughs
- **Frequency**: ν (nu) number of waves that pass a particular point in one second
- **Amplitude**: the vertical distance from the midline of waves to the top of the peak or the bottom of the trough
- Wave properties are mathematically related as:

$$c = \lambda \nu$$

where
- $c = 2.99792458 \times 10^8$ m/s (speed of light)
- λ = wavelength (in meters, m)
- ν = frequency (reciprocal seconds, s^{-1})

Source: Sapna Gupta

Solved Problem:
The wavelength of a laser pointer is reported to be 663 nm. What is the frequency of this light?

$$\upsilon = \frac{c}{\lambda}$$

$$\lambda = 663 \text{ nm} \times \frac{10^{-9} \text{m}}{\text{nm}} = 6.63 \times 10^{-7} \text{m}$$

$$\upsilon = \frac{3.00 \times 10^8 \text{ m/s}}{6.63 \times 10^{-7} \text{m}} = 4.52 \times 10^{14} \text{ s}^{-1}$$

Nature of Light

- In 1801, Thomas Young, a British physicist, showed that light could be diffracted; which is a wave property.

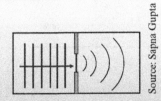

Source: Sapna Gupta

- The **photoelectric effect** was first observed by Heinrich Hertz 1887. It is the ejection of an electron from the surface of a metal or other material when light shines on it.
- The wave theory could not explain the photoelectric effect because this discovery means that light has energy also.

Quantum Theory

- 1900 - Max Planck
- Radiant energy could only be emitted or absorbed in discrete quantities
- Quantum: packets of energy
- Revolutionized way of thinking (energy is quantized)
- Energy of a single quantum of energy

$$E = h\nu$$

where

E = energy (in Joules)

h = Planck's constant 6.63×10^{-34} J · s

ν = frequency

Notes:

Photoelectric Effect

- Electrons ejected from a metal's surface when exposed to light of certain frequency
- Einstein proposed that particles of light are really **photons** (packets of light energy) and deduced that

$$E_{photon} = h\nu$$

- Only light with a frequency of photons such that $h\nu$ equals the energy that binds the electrons in the metal is sufficiently energetic to eject electrons. (*Threshold frequency*)
- If light of higher frequency is used, electrons will be ejected and will leave the metal with additional kinetic energy.
  – (what is the relationship between energy and frequency?)
- Light of at least the threshold frequency **and** of greater *intensity* will eject *more* electrons.

Notes:

Chapter 7: Quantum Theory

Solved Problem:
Calculate the energy (in joules) of a photon with a wavelength of 700.0 nm

$$\lambda = 700.0 \text{ nm} \times \frac{10^{-9} \text{ m}}{\text{nm}} = 7.00 \times 10^{-7} \text{ m}$$

Using $c = \lambda \nu$

$$\nu = \frac{3.00 \times 10^8 \text{ m/s}}{7.00 \times 10^{-7} \text{ m}} = 4.29 \times 10^{14} \text{ s}^{-1}$$

$E_{photon} = h\nu$

$$E = (6.63 \times 10^{-34} \text{ J} \cdot \text{s})(4.29 \times 10^{14} \text{ s}^{-1})$$

$$E = 2.84 \times 10^{-19} \text{ J}$$

Notes:

Bohr's Theory of the Atom

- Planck's theory along with Einstein's ideas not only explained the photoelectric effect, but also made it possible for scientists to unravel the idea of atomic line spectra
- *Line spectra*: emission of light only at specific wavelengths
- Every element has a unique emission spectrum
- Often referred to as "fingerprints" of the element

Source: Sapna Gupta

Notes:

Chapter 7: Quantum Theory 201

Line Spectrum

Line spectra of elements on the web. (need Java)

Line Spectrum of Hydrogen

- The *Rydberg equation*:

$$\frac{1}{\lambda} = R_\infty \left(\frac{1}{n_1^2} - \frac{1}{n_2^2} \right)$$

- Balmer (initially) and Rydberg (later) developed the equation to calculate all spectral lines in hydrogen
- Bohr's contribution: showed only valid energies for hydrogen's electron with the following equation

$$E_n = 2.18 \times 10^{-18} \, \text{J} \left(\frac{1}{n^2} \right)$$

- As the electron gets closer to the nucleus, E_n becomes larger in absolute value but also more negative.
- Ground state: the lowest energy state of an atom
- Excited state: each energy state in which $n > 1$
- Each spectral line corresponds to a specific transition
- Electrons moving from ground state to higher states require energy; an electron falling from a higher to a lower state releases energy
- Bohr's equation can be used to calculate the energy of these transitions within the H atom

Transition Between Different Levels

- An electron can change energy levels by absorbing energy to move to a higher energy level or by emitting energy to move to a lower energy level.
- For a hydrogen electron, the energy change is given by

$$\Delta E = E_f - E_i$$

$$\Delta E = -R_H \left(\frac{1}{n_f^2} - \frac{1}{n_i^2} \right) \quad R_H = 2.179 \times 10^{-8} \text{ J, Rydberg constant}$$

The energy of the emitted or absorbed photon is related to ΔE:

$$E_{photon} = |\Delta E_{electron}| = h\nu$$

h = Planck's constant

We can now combine these two equations: $h\nu = \left| -R_H \left(\frac{1}{n_f^2} - \frac{1}{n_i^2} \right) \right|$

Transition...contd.

Light is absorbed by an atom when the electron transition is from lower n to higher n ($n_f > n_i$). In this case, ΔE will be positive.

Light is emitted from an atom when the electron transition is from higher n to lower n ($n_f < n_i$). In this case, ΔE will be negative.

An electron is ejected when $n_f = \infty$.

Chapter 7: Quantum Theory

Solved Problem:
What is the wavelength of the light emitted when the electron in a hydrogen atom undergoes a transition from $n = 6$ to $n = 3$?

$n_i = 6$
$n_f = 3$
$R_H = 2.179 \times 10^{-18}$ J

$$\Delta E = -R_H \left(\frac{1}{n_f^2} - \frac{1}{n_i^2} \right)$$

$$|\Delta E| = \frac{hc}{\lambda} \quad \text{so} \quad \lambda = \frac{hc}{|\Delta E|}$$

$$\Delta E = \left(-2.179 \times 10^{-18} \text{ J} \right) \left(\frac{1}{3^2} - \frac{1}{6^2} \right) = -1.816 \times 10^{-19} \text{ J}$$

$$\lambda = \frac{\left(6.626 \times 10^{-34} \text{ J} \cdot \text{s} \right) \left(3.00 \times 10^8 \frac{\text{m}}{\text{s}} \right)}{\left| \left(-1.816 \times 10^{-19} \text{ J} \right) \right|} = 1.094 \times 10^{-6} \text{ m}$$

Notes:

Planck
Vibrating atoms have only certain energies:
$E = h\text{n}$ or $2h\text{n}$ or $3h\text{n}$

Einstein
Energy is quantized in particles called photons:
$E = h\text{n}$

Bohr
Electrons in atoms can have only certain values of energy. For hydrogen:

$$E = -\frac{R_H}{n^2}$$

$R_H = 2.179 \times 10^{-18}$ J, n = principal quantum number

Notes:

Chapter 7: Quantum Theory

- Light has properties of both waves and particles (matter). But **what** about matter?
- In 1923, **Louis de Broglie**, a French physicist, reasoned that particles (matter) might also have wave properties.
- The wavelength of a particle of mass, m (kg), and velocity, v (m/s), is given by the de Broglie relation:

$$\lambda = \frac{h}{mv}$$

$$h = 6.626 \times 10^{-34} \, J \cdot s$$

- Building on de Broglie's work, in 1926, **Erwin Schrödinger** devised a theory that could be used to explain the wave properties of electrons in atoms and molecules.
- In 1927, **Werner Heisenberg** showed how it is impossible to know with absolute precision both the position, x, and the momentum, p, of a particle such as electron.
- Because $p = mv$ this uncertainty becomes more significant as the mass of the particle becomes smaller.

$$(\Delta x)(\Delta p) \geq \frac{h}{4\pi}$$

- Solving Schrödinger's equation gives us a **wave function**, represented by the Greek letter psi, ψ, which gives information about a particle in a given energy level.
- Psi-squared, ψ^2, gives us the probability of finding the particle in a region of space.

Key Points

- Electromagnetic spectrum
- Wavelength, frequency, energy (calculate)
- Quanta (of light - photon)
- Photoelectric effect
- Emission spectra
- Ground state vs excited state
- Heisenberg uncertainty principle
- Niels Bohr's line spectrum
- Schrodingers wave function

Ch 7/ PowerPoint Study- 1 Quantum Theory   Name: _____

Answer these questions as you are watching the videos. They are due in class.
These questions are not just for you to answer but also to prepare you for the exam.
<u>*Make sure you understand what you are writing and not just copy from the text book.*</u> **Show all work.**

1. Arrange the following radiations in increasing wavelength.

    Ultra-violet, radio, X-rays, visible

2. True or False: wavelength is directly proportional to energy.

3. True or False: frequency is directly proportional to energy.

4. Write briefly the contribution of the following scientists:

    Max Planck:

    Bohr:

    Schrodinger:

    Heisengberg:

5. Explain the following terms:

    Quantum:

    Wave function:

    Ground state:

Chapter 7 Atomic Structure -2 Quantum Numbers

Dr. Sapna Gupta

Quantum Numbers

According to quantum mechanics, each electron is described by four quantum numbers:

1. Principal quantum number (n)
2. Angular momentum quantum number (l)
3. Magnetic quantum number (m_l)
4. Electron spin quantum number (m_s)

The first three define the wave function for a particular electron. The fourth quantum number refers to the magnetic property of electrons.

A wave function for an electron in an atom is called an atomic orbital (described by three quantum numbers—n, l, m_l).

It describes a region of space with a definite shape where there is a high probability of finding the electron.

Notes:

- **Principal quantum number (n)** - designates size of the orbital
- Integer values: 1, 2, 3, and so forth
- The larger the "n" value, the greater the average distance from the nucleus
- Correspond to quantum numbers in Bohr's model

- **Angular momentum quantum number (l)** - shape of the atomic orbital
- Integer values: 0 to $n-1$
- 0 = s sublevel;
  1 = p
  2 = d
  3 = f

- **Magnetic quantum number (m_l)** - orientation of the orbital in space (think in terms of x, y and z axes)
- Integer values: $-l$ to 0 to $+l$
- There are 2e- in each orientation

- **Electron spin quantum number (m_s)** - describes the spin of an electron that occupies a particular orbital
- Values: +1/2 or -1/2
- Electrons will spin opposite each other in the same orbital

Allowed Values

When n is	l can be	When l is	m_l can be
1	Only 0	0	Only 0
2	0 or 1	0	Only 0
		1	-1, 0, +1
3	0, 1 or 2	0	Only 0
		1	-1, 0, +1
		2	-2, -1, 0, +1, +2
4	0, 1, 2 or 3	0	Only 0
		1	-1, 0, +1
		2	-2, -1, 0, +1, +2
		3	-3, -2, -1, 0, +1, +2, +3

Notes:

- When $n = 1$, l has only one value, 0.
- When $l = 0$, m_l has only one value, 0.

So the first shell ($n = 1$) has one subshell, an *s*-subshell, 1*s*. That subshell, in turn, has one orbital; 2e⁻.

- When $n = 2$, l has two values, 0 and 1.
- When $l = 0$, m_l has only one value, 0. So there is a 2*s* subshell with one orbital; 2e⁻.
- When $l = 1$, m_l has only three values, -1, 0, 1. So there is a 2*p* subshell with three orbitals; 6e⁻.

- When $n = 3$, l has three values, 0, 1, and 2.
- When $l = 0$, m_l has only one value, 0. So there is a 3*s* subshell with one orbital; 2e⁻.
- When $l = 1$, m_l has only three values, -1, 0, 1. So there is a 3*p* subshell with three orbitals, 6e⁻.
- When $l = 2$, m_l has only five values, -2, -1, 0, 1, 2. So there is a 3*d* subshell with five orbitals; 10e⁻.

Notes:

Atomic View of Quantum Numbers

$n = 1$
$l = n-1 = 0$ (s)
$m_l = 0$
e⁻ = 2

$n = 2$
$l = n-1 = 0, 1$ (s,p)
$m_l = 0$
 -1, 0, +1
e⁻ = 2+6

$n = 3$
$l = n-1 = 0, 1, 2$ (s,p,d)
$m_l = 0$
 -1, 0, +1
 -2, -1, 0, +1, +2
e⁻ = 2+6+10

$n = 4$
$l = n-1 = 0, 1, 2, 3$ (s,p,d,f)
$m_l = 0$
 -1, 0, +1
 -2, -1, 0, +1, +2
 -3, -2, -1, 0, +1, +2, +3
e⁻ = 2+6+10+14

Notes:

Solved Problem:
Which of the following are permissible sets of quantum numbers?

$n = 4, l = 4, m_l = 0, m_s = ½$
$n = 3, l = 2, m_l = 1, m_s = -½$
$n = 2, l = 0, m_l = 0, m_s = ^3/2$
$n = 5, l = 3, m_l = -3, m_s = ½$

(a) Not permitted. When $n = 4$, the maximum value of l is 3.
(b) Permitted.
(c) Not permitted; m_s can only be +½ or –½.
(b) Permitted.

Shapes of Atomic Orbitals

• An *s* orbital is spherical.

• A *p* orbital has two lobes along a straight line through the nucleus, with one lobe on either side.

• A *d* orbital has a more complicated shape.

Chapter 7: Quantum Theory

s - orbital

The blue-green cross-sectional view of a 1s orbital and a 2s orbital highlights the difference in the two orbitals' sizes.

The larger sizes of the s orbitals show a better sense of them in three dimensions.

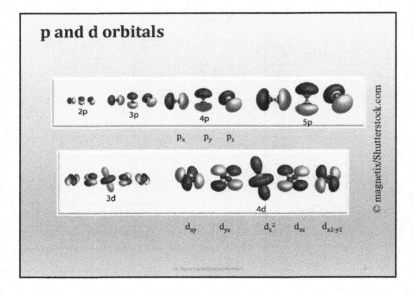

Notes:

p and d orbitals

p_x p_y p_z

d_{xy} d_{yz} $d_z{}^2$ d_{xz} d_{x2-y2}

Notes:

Chapter 7: Quantum Theory 211

F-orbitals

4f

© magnetix/Shutterstock.com

Notes:

All orbitals – 4th Shell

4s 2 e-
4p 6 e-
4d 10 e-
4f 14 e-

© magnetix/Shutterstock.com

Notes:

Chapter 7: Quantum Theory

Key Points

- Quantum numbers (n, l, m_l, m_s) predict values and possible sets
- Shapes of orbitals

Notes:

Ch 7/ PowerPoint Study-Quantum Numbers

Name: _____

Answer these questions as you are watching the videos. They are due in class.
These questions are not just for you to answer but also to prepare you for the exam.
Make sure you understand what you are writing and not just copy from the text book. **Show all work.**

1. What is the principal quantum number and what aspect of the atom does it represent?

2. How is angular momentum number calculated? What is/are the value(s) of l, when n = 4?

3. Give the different value(s) of magnetic quantum numbers that is/are possible when $l = 1$? What do these numbers mean? How many total electrons are possible when $l = 1$?

4. What does a +1/2 mean for an electron's magnetic spin number?

5. What alphabet of the angular quantum number associated with $l = 2$? (i.e. spdf?) _____

6. If n = 3, can $l = 2$? _____

7. If n = 4, can $m_l = +5$? _____

8. The quantum numbers for 3d are: n=3, l=1, m_l=1, m_s=+1/2? If not, write the correct numbers.

Ch 7 Worksheet/Quantum Numbers   Name: _____

1. Which type of electromagnetic radiation has the shortest wavelength?
   A. red light
   B. x rays
   C. microwaves
   D. gamma rays
   E. blue light

2. What is the wavelength of a photon having a frequency of 4.50×10^{14} Hz? ($c = 3.00 \times 10^8$ m/s)
   A. 667 nm
   B. 1.50×10^{-3} nm
   C. 4.42×10^{-31} nm
   D. 0.0895 nm
   E. 2.98×10^{-10} nm

3. What is the energy of a photon of electromagnetic radiation with a wavelength of 877.4 nm?
   ($c = 3.00 \times 10^8$ m/s, $h = 6.63 \times 10^{-34}$ J·s)
   A. 2.16×10^{-19} J
   B. 5.82×10^{-40} J
   C. 2.16×10^{-28} J
   D. 3.42×10^{14} J
   E. 1.94×10^{-39} J

4. Which of the following is/are correct postulates of Bohr's theory of the hydrogen atom?

   1. The energy of an electron in an atom is quantized (i.e. only specific energy values are possible).

   2. The principal quantum number (n), specifies each unique energy level.

   3. An electron transition from a lower energy level to a higher energy level results in an emission of a photon of light.
      A. 1 only
      B. 2 only
      C. 3 only
      D. 1 and 2
      E. 1, 2, and 3

5. List all the orbitals when n = 4.

6. Give the formula that relates the number of possible values of m_l to the value of l.

7. Which of the following subshells cannot exist: (a) 1p; (b) 4f; (c) 2d; (d) 5p; (e) 3f? Why not?

8. List all possible values of m_l for each of the indicated subshells. What role does the principal quantum number n play in determining your answer?

Subshell	Values of m_l
(a) 4s	
(b) 2p	
(c) 3d	
(d) 5f	

9. Which of the following sets of quantum numbers (n, l, ml, ms) refers to a 3d orbital?
   A. 2 1 0 + 1/2
   B. 5 4 3 + 1/2
   C. 4 2 1 − 1/2
   D. 4 3 1 − 1/2
   E. 3 2 1 − 1/2

10. An orbital with the quantum numbers n = 3, l = 1, m_l = −1, may be found in which subshell?
    A. 3f
    B. 3d
    C. 3p
    D. 3g
    E. 3s

Chapter 8

Electronic Configurations, Element Properties and the Periodic Table

Filling Electrons (spdf notation and box configurations)

1. <u>Aufbau principle</u>: building up of atom from bottom up i.e. electrons occupy orbitals of the lowest energy available.
2. <u>Pauli's exclusion principle</u>: no two electrons in an atom have the same quantum number, i.e. an atomic orbital that has two electrons must have the opposite spins (magnetic spin number)
3. <u>Hund's Rule</u>: electrons are filled in the empty orbital first before pairing up. The singly filled electrons have the same magnetic spin.

- Electronic configurations can be also filled as noble gas configuration. Select the noble gas just before the element.
- Transition metal electrons (d) go in the penultimate core shell.
- Periodic table has s, p, d and f blocks
- Electronic configuration of ions:
  - <u>Anions</u>: add electrons in the p orbitals or d in case of expanded shell.
  - <u>Cations</u>: for main group remove electrons from s or p. For transition metals remove valence s electrons first and then the d.

Periodic Properties

1. <u>Radius</u>:
   a. Atomic radii: decreases across and increases down the PT
   b. Ionic radii: cations smaller than original atoms while anions larger than original atom.
2. <u>Ionization Energy</u> (IE): energy required to remove an electron from ground state atom or ion in a gas state. Increases across and decreases down the PT.
3. <u>Electron Affinity</u>: energy change associated with an atom when it gains an electron in a gas state. More energy is released across the PT
4. <u>Metallic Character</u>: increases top to bottom in a group and decreases from left to right within a period.
5. Also compare – ionic size vs atomic size.

Properties of elements according to groups

Group I – low IE; all are M^+ ions; react with oxygen and water readily to give oxides and hydroxides.
Group II – less reactive than group I; reactions with water varies; they still give hydroxides. Free metals react with acids to give hydrogen gas.
Group III – Al forms an oxide coating in presence of oxygen.
Group IV – has non metals and metals. Don't react with water but with acids to give hydrogen gas.
Group V – has non metals and metals. Oxides of elements will react with water to form acids.
Group VI – most are non metals. Non metal oxides of elements will react with water to form acids.
Group VII – high IE. Form mostly ionic compounds with metals; hydrogen halides are acids and electrolytes.
Group VIII – noble elements – unreactive.

Key words:

Metals, Nonmetals, Metalloids and Noble Gases	Isoelectric	Allotropes
Effective nuclear charge	Valence electrons	Core electron

Non metal oxides + water = acids;   Metal oxides + water = bases

Chapter 8 Electronic Configurations

Dr. Sapna Gupta

Structure of Atom

- We have learned that electrons exist in atoms in specific locations and are always in motion – in the orbit and within the orbital (magnetic spin – m_s).
- In this chapter we will learn how to fill these electrons in atoms in two different ways:
  - Electronic configuration (spdf notation): here we fill electrons in the various sub shells according to set rules.
  - Orbital diagram (box configuration): in this case we show electrons as arrows and subshells as boxes and then fill out the electrons.

The Principles of Filling Electrons

- <u>Pauli's exclusion principle</u>: no two electrons can have the same quantum numbers.
- <u>Aufbau principle</u>: building up the electronic configuration using ground state energies. Electrons and filled in the subshells according to energy levels of shells and subshells.
- <u>Hund's rule</u>: Fill the electrons in the subshells singly first, then pair them up. (magnetic quantum number – m_l)

Chapter 8: Electronic Configurations, Element Properties and the Periodic Table

The First Three Rows

Row 1

H	$1s^1$
He	$1s^2$

Row 2

Li	$1s^2\ 2s^1$
Be	$1s^2\ 2s^2$
B	$1s^2\ 2s^2\ 2p^1$
C	$1s^2\ 2s^2\ 2p^2$
N	$1s^2\ 2s^2\ 2p^3$
O	$1s^2\ 2s^2\ 2p^4$
F	$1s^2\ 2s^2\ 2p^5$
Ne	$1s^2\ 2s^2\ 2p^6$

Row 3

Na	$1s^2\ 2s^2\ 2p^6\ 3s^1$
Mg	$1s^2\ 2s^2\ 2p^6\ 3s^2$
Al	$1s^2\ 2s^2\ 2p^6\ 3s^2\ 3p^1$
Si	$1s^2\ 2s^2\ 2p^6\ 3s^2\ 3p^2$
P	$1s^2\ 2s^2\ 2p^6\ 3s^2\ 3p^3$
S	$1s^2\ 2s^2\ 2p^6\ 3s^2\ 3p^4$
Cl	$1s^2\ 2s^2\ 2p^6\ 3s^2\ 3p^5$
Ar	$1s^2\ 2s^2\ 2p^6\ 3s^2\ 3p^6$

Rules for Writing Electron Configurations

- Electrons reside in orbitals of lowest possible energy
- Maximum of 2 electrons per orbital
- Electrons do not pair in degenerate orbitals (same energy orbitals) if an empty orbital is available
- Orbitals fill in the following order:
  1s 2s 2p 3s 3p 4s 3d 4p 5s 4d 5p 6s
  or as shown on the right.
  (Follow the arrow)

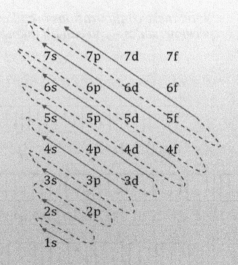

Box Configuration

To fill the electrons in box use full or half arrows to show electrons (as shown for He). Electrons are filled singly first and then paired up.

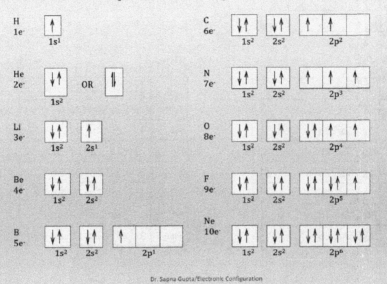

Some More Examples

- *Note: these configurations are filled with full arrows, unlike the previous slides – you can fill them either way.*

- $Z = 20$ (Ca) $1s^2 2s^2 2p^6 3s^2 3p^6 4s^2$

- $Z = 35$ (Br) $1s^2 2s^2 2p^6 3s^2 3p^6 4s^2 3d^{10} 4p^5$

- $Z = 26$ (Fe) $1s^2 2s^2 2p^6 3s^2 3p^6 4s^2 3d^6$

Chapter 8: Electronic Configurations, Element Properties and the Periodic Table

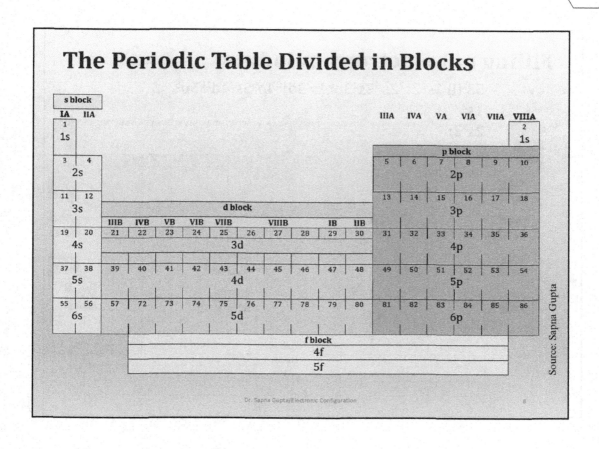

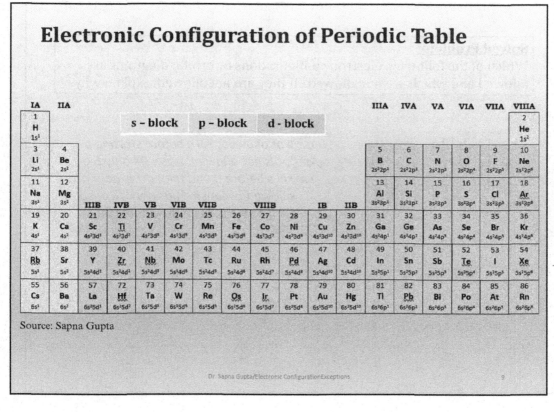

Chapter 8: Electronic Configurations, Element Properties and the Periodic Table

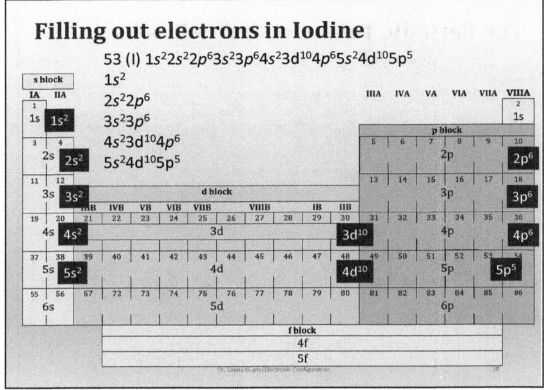

Solved Problem:
Which of the following electron configurations or orbital diagrams are allowed and which are not allowed? If they are not allowed, explain why?

a. $1s^2 2s^1 2p^3$

b. $1s^2 2s^2 2p^8$

c. $1s^2 2s^2 2p^6 3s^2 3p^6 3d^{11}$

a. Not allowed; Fill s before starting p.

b. p^8 is not allowed. p can fill only 6 e-

c. Fill s before d and then d^{11} is not allowed because d has only 10 e- so fill it to 9 e (which also should be $s^1 d^{10}$)

Chapter 8: Electronic Configurations, Element Properties and the Periodic Table

Key Words

- Spdf and box configurations
- Pauli exclusion principle
- Hund's rule
- Aufbau principle

Ch 8/ PowerPoint Study-1 Electronic Configuration   Name: _____

Answer these questions as you are watching the videos. They are due in class.
These questions are not just for you to answer but also to prepare you for the exam.
Make sure you understand what you are writing and not just copy from the text book. **Show all work.**

1. Name the element with the following electronic configurations.

	$1s^2\ 2s^2\ 2p^6\ 3s^2\ 3p^6\ 4s^2\ 3d^{10}\ 4p^1$
	$1s^2\ 2s^2\ 2p^6\ 3s^2\ 3p^6\ 4s^2\ 3d^{10}\ 4p^6\ 5s^2\ 4d^{10}\ 5p^5$
	↓↑ | ↓↑ | ↓↑ ↓↑ ↓↑ | ↑
	↓↑ | ↓↑ | ↓↑ ↓↑ ↓↑ | ↓↑ | ↓↑ ↓↑ ↓↑ | ↓↑ | ↑ ↑ ↑
	[Ne] $3s^2\ 3p^3$
	[Ar] $4s^2\ 3d^6$

2. Answer the following questions for fluorine.

    a. How many total electrons does fluorine have? _____

    b. Write the electronic configuration of fluorine. _____

    c. How many "total" s electrons? _____

    d. Which "block" does fluorine belong to? _____

    e. Which orbital has the last electron, spdf? _____

    f. Write the box configuration for fluorine below.

3. Answer the questions below for calcium.

    a. How many total electrons does calcium have? _____

    b. Write the electronic configuration of calcium. _____

    c. How many electrons are in the last p orbital of calcium? _____

    d. Which "block" does calcium belong in the periodic table? _____

    e. Write the box configuration for calcium below.

Chapter 8: Electronic Configurations, Element Properties and the Periodic Table

4. Explain clearly what is wrong with the following configurations by writing which principle (Aufbau, Pauli or Hund's) is being violated. Give the correct answer and write a sentence to explain your answer. (NOTE: Make sure the electrons in the element are represented in the configuration given).

Element	
a) P:	$1s^2\ 2s^2\ 2p^6\ 3s^3\ 3p^3$
b) Ti:	$1s^2\ 2s^2\ 2p^6\ 3s^2\ 3p^6\ 3d^2$
c) Si	↓↑ ↓↑ ↓↑ ↓↑ ↓↑ ↑ ↑ ↑ ↑
d) Mg	↓↑ ↓↑ ↓↑ ↓↑ ☐ ↓↑ ↓↑ ☐

Chapter 8 Electronic Configurations – 2 Exceptions, Properties, Ionic Configruations, etc.

Dr. Sapna Gupta

Exceptions

- There are a things to watch out for when filling out shells.
  - Fill s in the higher n number before starting d:
    - Fill the 4s before 3d because of energy consideration. Same goes for 5s before 4d.
  - Fill s completely before d for two columns
    - Chromium (4th column in transition metals) and elements below: Should be $4s^2, 3d^4$; But is **$4s^1, 3d^5$** (for Mo: $5s^1, 4d^5$) This is to make the d configuration more stable.
    - Copper and the elements below it are filled as: Should be $4s^2\ 3d^9$; But is **$4s^1, 3d^{10}$** (for Ag: $5s^2, 4d^{10}$)

Electronic Configuration of Periodic Table

IA	IIA											IIIA	IVA	VA	VIA	VIIA	VIIIA
1 H $1s^1$		s – block			p – block		d – block										2 He $1s^2$
3 Li $2s^1$	4 Be $2s^2$											5 B $2s^22p^1$	6 C $2s^22p^2$	7 N $2s^22p^3$	8 O $2s^22p^4$	9 F $2s^22p^5$	10 Ne $2s^22p^6$
11 Na $3s^1$	12 Mg $3s^2$	IIIB	IVB	VB	VIB	VIIB	VIIIB			IB	IIB	13 Al $3s^23p^1$	14 Si $3s^23p^2$	15 P $3s^23p^3$	16 S $3s^23p^4$	17 Cl $3s^23p^5$	18 Ar $3s^23p^6$
19 K $4s^1$	20 Ca $4s^2$	21 Sc $4s^23d^1$	22 Ti $4s^23d^2$	23 V $4s^23d^3$	24 Cr $4s^13d^5$	25 Mn $4s^23d^5$	26 Fe $4s^23d^6$	27 Co $4s^23d^7$	28 Ni $4s^23d^8$	29 Cu $4s^13d^{10}$	30 Zn $4s^23d^{10}$	31 Ga $4s^24p^1$	32 Ge $4s^24p^2$	33 As $4s^24p^3$	34 Se $4s^24p^4$	35 Br $4s^24p^5$	36 Kr $4s^24p^6$
37 Rb $5s^1$	38 Sr $5s^2$	39 Y $5s^24d^1$	40 Zr $5s^24d^2$	41 Nb $5s^14d^4$	42 Mo $5s^14d^5$	43 Tc $5s^24d^5$	44 Ru $5s^14d^7$	45 Rh $5s^14d^8$	46 Pd $4d^{10}$	47 Ag $5s^14d^{10}$	48 Cd $5s^24d^{10}$	49 In $5s^25p^1$	50 Sn $5s^25p^2$	51 Sb $5s^25p^3$	52 Te $5s^25p^4$	53 I $5s^25p^5$	54 Xe $5s^25p^6$
55 Cs $6s^1$	56 Ba $6s^2$	57 La $6s^25d^1$	72 Hf $6s^25d^2$	73 Ta $6s^25d^3$	74 W $6s^25d^4$	75 Re $6s^25d^5$	76 Os $6s^25d^6$	77 Ir $6s^25d^7$	78 Pt $6s^15d^9$	79 Au $6s^15d^{10}$	80 Hg $6s^25d^{10}$	81 Tl $6s^26p^1$	82 Pb $6s^26p^2$	83 Bi $6s^26p^3$	84 Po $6s^26p^4$	85 At $6s^26p^5$	86 Rn $6s^26p^6$

Source: Sapna Gupta

Valence and Core Electrons

- Silicon has 4 valence electrons (those in the $n = 3$ principal shell) and 10 core electrons.

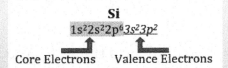

- Selenium has 6 valence electrons (those in the $n = 4$ principal shell). All other electrons, including those in the $3d$ orbitals, are core electrons.

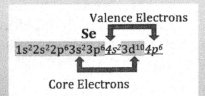

Noble Gas Configurations

- This helps to shorten the electronic configurations so we don't have to write long notations.
- Take the noble gas of the previous period and continue on filling with the rest of the electrons.
- E.g. Bromine configuration is:

$$1s^2 2s^2 2p^6 3s^2 3p^6 4s^2 3d^{10} 4p^5$$

- But the noble gas configuration is:
$$[Ar]4s^2 3d^{10} 4p^5$$

(It is important to remember that the "d" electrons are core so the only the s and p electrons are valence electrons)

Configurations for Ions

- Cations: remove the valence shell electrons of the higher energy first.

  Al: $1s^2 2s^2 2p^6 3s^2 3p^1$

  Al^{3+}: $1s^2 2s^2 2p^6$

- Anions: add the electrons to the lower subshell of valence shell.

  S: $1s^2 2s^2 2p^6 3s^2 3p^4$

  S^{2-}: $1s^2 2s^2 2p^6 3s^2 3p^6$

Paramagnetism and Diamagnetism

A **paramagnetic substance** is one that is weakly attracted by a magnetic field, usually as the result of *unpaired electrons*.

A **diamagnetic substance** is not attracted by a magnetic field generally because it has *only paired electrons*.

Visit UC Davis ChemWiki for more information.

Key Words

- Noble gas configuration
- Valence and core electrons
- Electronic configuration of ions
- Paramagnetism
- Diamagnetism

Ch 8/ PowerPoint Study-2 Electronic Configuration Exceptions

Name: _____

Answer these questions as you are watching the videos. They are due in class.
These questions are not just for you to answer but also to prepare you for the exam.
Make sure you understand what you are writing and not just copy from the text book. **Show all work.**

1. How many core and valence electrons are in the following elements?

                    calcium   sulfur   iron

   Core e^-

   Valence e^-

2. What is the noble gas can be used for the electronic configuration of the following elements?

                    rubidium (Rb)   chromium   sulfur

   Noble Gas:

3. Write the spdf and box configuration of the following ions. Do not use noble gas notation.

   a. Al^{3+}

   b. O^{2-}

Chapter 8 Periodic Properties

Dr. Sapna Gupta

Periodicity

- Initially the periodic table had been arranged by atomic weight but was later changed by Mendeleev to atomic numbers.
- When organizing these elements Mendeleev found that elements in each group had the same chemical properties.
- There was also a trend in the rows and within the groups of the periodic table. This is periodicity.
- There are three main trends to study:
  - Atomic radii
  - Ionization energy and
  - Electron affinity
- Effective Nuclear Charge – all the above three trends are based on the fundamental understanding of effective nuclear charge.

Effective Nuclear Charge (Z_{eff})

- This is the attraction of negative electrons to the positive nucleus.
- All periodic trends are based on Z_{eff}
- As electrons are added to the same shell the Z_{eff} increases
- As electrons are added to a shell farther away from the nucleus the Z_{eff} decreases.

	IA	IIA	IIIA	IVA	VA	VIA	VIIA	VIIIA
	1							2
1s	H							He
	1.0							1.34
	3	4	5	6	7	8	9	10
2s, 2p	Li	Be	B	C	N	O	F	Ne
	1.26	1.58	1.56	1.82	2.07	2.07	2.26	2.52
	11	12	13	14	15	16	17	18
3s, 3p	Na	Mg	Al	Si	P	S	Cl	Ar
	1.64	2.25						

Atomic Radii; Ionization Energy; Electron Affinity; Metallic Character

- Atomic radius is half the distance between two nuclei.
- Ionization Energy is the amount of energy required to remove one electron from an atom in a gas phase.
- Electron Affinity is the measure of energy change when an electron is added to a valence shell. A negative energy measurement means an anion is formed and positive energy measurement means cation is formed.
- Metallic Character is how much a metal behaves like a metal e.g. conducts electricity and heat, malleability, reactivity with acids etc.

Atomic Radii

Atomic radius is half the distance between two nuclei.

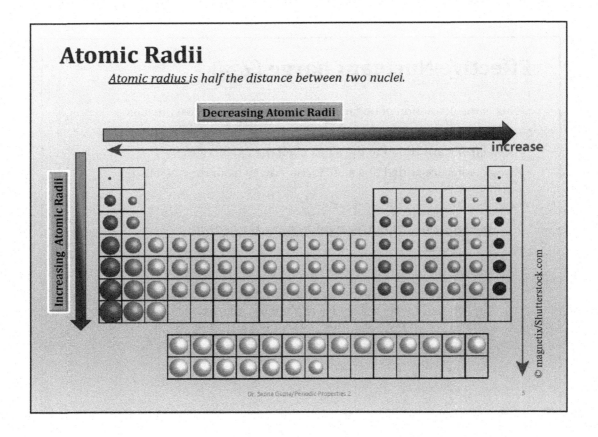

Chapter 8: Electronic Configurations, Element Properties and the Periodic Table

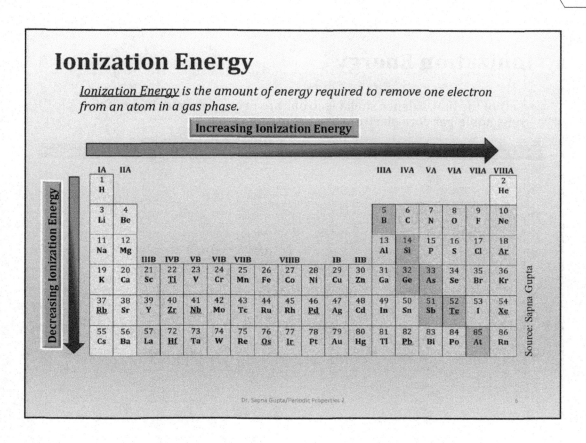

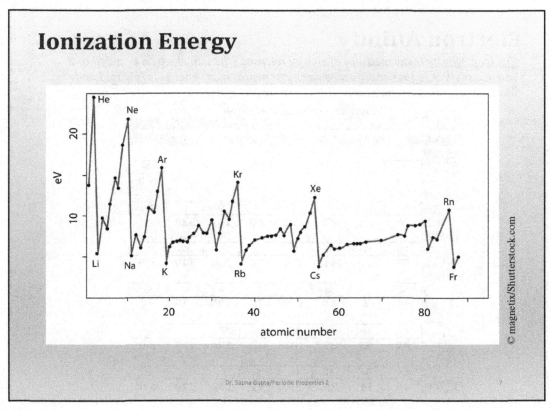

Ionization Energy

- Left of the line, valence shell electrons are being removed. Right of the line, noble-gas core electrons are being removed.

Element	First	Second	Third	Fourth	Fifth	Sixth	Seventh
H	1312						
He	2372	5250					
Li	520	7298	11,815				
Be	900	1757	14,848	21,006			
B	801	2427	3660	25,026	32,827		
C	1086	2353	4620	6223	37,831	47,277	
N	1402	2856	4578	7475	9445	53,326	64,360
O	1314	3388	5300	7469	10,990	13,326	71,330
F	1681	3374	6050	8408	11,023	15,164	17,868
Ne	2081	3952	6122	9371	12,177	15,238	19,999

Electron Affinity

Electron Affinity is the measure of energy released when an electron is added to a valence shell. A higher energy measurement means more energy is released and lower energy measurement means less energy is released.

Increasing Electron Affinity →
Decreasing Electron Affinity ↓

	IA	IIA	IIIA	IVA	VA	VIA	VIIA	VIIIA
Row 1	H 73							He ≤0
Row 2	Li 60	Be ≤0	B 27	C 122	N ≤0	O 141	F 328	Ne ≤0
Row 3	Na 53	Mg ≤0	Al 44	Si 134	P 72	S 200	Cl 349	Ar ≤0
Row 4	K 48	Ca 2	Ga 41	Ge 119	As 78	Se 195	Br 325	Kr ≤0
Row 5	Rb 47	Sr 5	In 37	Sn 107	Sb 101	Te 190	I 295	Xe ≤0
Row 6	Cs 46	Ba 14	Tl 36	Pb 35	Bi 91	Po 180	At 270	Rn ≤0

Metallic Character

Metallic Character	Nonmetals	Metalloids
•Shiny, lustrous, malleable •Good conductors •Low *IE* (form cations) •Form ionic compounds with chlorine •Form basic, ionic compounds with oxygen •Metallic character increases top to bottom in group and decreases left to right across a period	•Vary in color, not shiny •Brittle •Poor conductors •Form acidic, molecular compounds with oxygen •High *EA* (form anions) •Group VII and VIII are all non metals	•Properties both of metals and nonmetals

Trends

Trend	Definition	Across a Group	Down a Group
Effective nuclear charge	This is the attraction of negative electrons to the positive nucleus	Increases Because electrons are added to the same shell	Decreases Because electrons are added to a new shell
Atomic radii	Atomic radius is half the distance between two nuclei	Decreases Because electrons are added to the same shell	Increases Electrons are added to a new shell
Ionization energy	IE is the amount of energy required to eject one electron from an atom in a gas phase	Increases Z increases, so harder to remove an electron	Decreases Z decreases so easy to remove an electron
Electron affinity	Electron Affinity: the measure of energy released when an electron is added to a valence shell	Increases easier to gain electrons	Decreases harder to add electrons
Metallic Character	how much a metal behaves like a metal e.g. conducts electricity and heat, malleability etc.	Decreases Z increases so harder to lose electrons	Increases Z decreases so easier to lose electrons

Chapter 8: Electronic Configurations, Element Properties and the Periodic Table

Ionic Radii

- Cations give electrons: so protons > electrons radii gets smaller
- Anions accept electrons: so protons < electrons radii gets larger

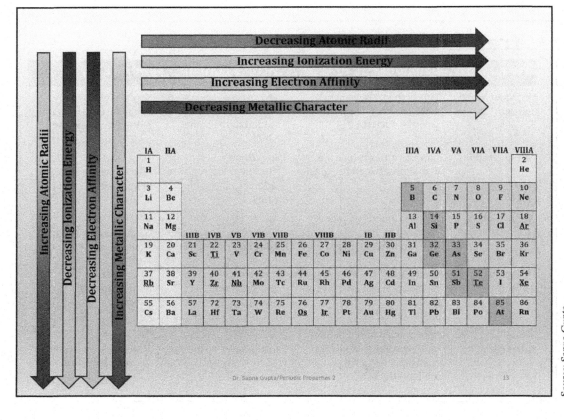

Chapter 8: Electronic Configurations, Element Properties and the Periodic Table

Solved Problem:
Refer to a periodic table and arrange the following elements in
a) order of increasing atomic radius: Br, Se, Te.

34 Se	35 Br
52 Te	

Te is larger than Se.
Se is larger than Br.

Br < Se < Te

b) order of increasing ionization energy: As, Br, Sb.

33 As		35 Br
51 Sb		

Sb is larger than As.
As is larger than Br.

Ionization energies:
Sb < As < Br

Isoelectric Elements/Ions

- Two or more species having the same electron configuration but different nuclear charges
- Atomic/ionic size varies significantly

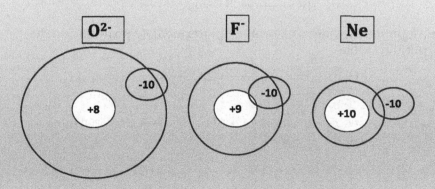

Properties of the Groups

- Groups I and II – all metals
- Groups VII and VIII – all non metals
- Transition metals – all metals
- Groups III, IV, V and VI – has non metals, metalloids and metals.

- Oxides: dissolving oxides in water give bases and acids.
  - Metal oxides give bases
  - Non metal oxides give acids
  - Amphoteric oxides can give both acids and bases

Reactions of Oxides

- Metal oxides are usually basic

$$Na_2O(s) + H_2O(l) \longrightarrow 2NaOH(aq)$$

- Nonmetal oxides are usually acidic

$$SO_3(g) + H_2O(l) \longrightarrow 2H_2SO_4(aq)$$

- Amphoteric oxides are located at intermediate positions on the periodic table

$$Al_2O_3(s) + 6HCl(aq) \longrightarrow 2AlCl_3(aq) + 3H_2O(l)$$

$$Al_2O_3(s) + 2NaOH(aq) + 3H_2O(l) \longrightarrow 2NaAl(OH)_4(aq)$$

Group I – Alkali Metals (ns^1)

- All metals
- Reactivity increases down the group
- Never found in nature in elemental state
- Oxides are M$_2$O
- All metals very reactive – have to be stored in special containers.
- All metals are very soft (unlike typical metals)

Group II – Alkaline Earth Metals (ns^2)

- These elements are metals
- Reactivity increases down the group.
- Less reactive than group I
- The oxides have the formula MO
- Some react with acids to form hydrogen gas

Group III ($ns^2\ np^1$)

Group III

5 **B** $2s^2 2p^1$	
13 **Al** $3s^2 3p^1$	
31 **Ga** $4s^2 4p^1$	
49 **In** $5s^2 5p^1$	
81 **Tl** $6s^2 6p^1$	

- Metalloid (B) and metals (all others)
- Al forms Al_2O_3 with oxygen and is an amphoteric oxide
- Al reacts with acid
- Other metals form +1 and +3

Group IV ($ns^2\ np^2$)

Group IV

6 **C** $2s^2 2p^2$
14 **Si** $3s^2 3p^2$
32 **Ge** $4s^2 4p^2$
50 **Sn** $5s^2 5p^2$
82 **Pb** $6s^2 6p^2$

- Nonmetal (C) metalloids (Si, Ge) and other metals
- Form +2 and +4 oxidation states
- Sn, Pb react with acid to produce H_2
- CO_2, SiO_2, and GeO_2 are acidic (decreasingly so).
- SnO_2 and PbO_2 are amphoteric.

Chapter 8: Electronic Configurations, Element Properties and the Periodic Table

Group V (ns² np³)

Group V

7	N	$2s^2 2p^3$
15	P	$3s^2 3p^3$
33	As	$4s^2 4p^3$
51	Sb	$5s^2 5p^3$
83	Bi	$6s^2 6p^3$

- Nonmetal (N_2, P) metalloid (As, Sb) and metal (Bi)
- Nitrogen, N_2 forms variety of oxides
- Phosphorus, P_4
- As, Sb, Bi (crystalline)
- HNO_3 and H_3PO_4 important industrially
- Nitrogen, phosphorus, and arsenic oxides are acidic.
- Antimony oxides are amphoteric.
- Bismuth oxide is basic.

Group VI (ns² np⁴)

Group VI

8	O	$2s^2 2p^4$
16	S	$3s^2 3p^4$
34	Se	$4s^2 4p^4$
52	Te	$5s^2 5p^4$
84	Po	$6s^2 6p^4$

- Nonmetals (O, S, Se)
- Metalloids (Te, Po)
- Oxygen, O_2
- Sulfur, S_8
- Selenium, Se_8
- Te, Po (crystalline)
- Oxides form acids in water; SO_2, SO_3, H_2S, H_2SO_4

Group VII – Halogens ($ns^2\ np^5$)

Group VII

9 **F** $2s^22p^5$	
17 **Cl** $3s^23p^5$	
35 **Br** $4s^24p^5$	
53 **I** $5s^25p^5$	
85 **At** $6s^26p^5$	

- All non metals
- All elements are diatomic molecules
- Gain one electron during reactions
- Physical property trend is gas (F_2 and Cl_2), liquid (Br_2), solid (I_2).
- Halogens form acids (HX, where X is a halogen)

Group VIII – Noble Gases ($ns^2\ np^6$)

Group VIII

2 **He** $1s^2$	
10 **Ne** $2s^22p^6$	
18 **Ar** $3s^23p^6$	
36 **Kr** $4s^24p^6$	
54 **Xe** $5s^25p^6$	
86 **Rn** $6s^26p^6$	

- All monatomic
- Filled valence shells
- All are gases
- Considered "inert" until 1963 when Xe and Kr were used to form compounds
- No major commercial use

Chapter 8: Electronic Configurations, Element Properties and the Periodic Table

Key Words

- Effective nuclear charge
- Atomic radii
- Ionization energy
- Electron affinity
- Metallic character
- Ionic radii
- Isoelectric elements/ions
- Basic oxide
- Acidic oxide

Ch 8/ PowerPoint Study-3 Electronic Configuration-Periodic Properties

Name: _____

Answer these questions as you are watching the videos. They are due in class.
These questions are not just for you to answer but also to prepare you for the exam.
Make sure you understand what you are writing and not just copy from the text book. **Show all work.**

1. Explain the following terms:
    a. Ionization energy:

    b. Electron affinity:

2. Circle the element with the
    a. higher ionization energy: Li or Be   Na or K   S or Br

    b. higher electron affinity: Li or Be   Na or K   S or Br

    c. larger cationic radii: Li^+ or Be^{2+} Na^+ or K^+

    d. larger anionic radii: S^{2-} or Cl^- N^{3-} or P^{3-}

3. Identify the ions/elements that are isoelectronic and write how many electrons the pairs have.
    Li^+, Be^{2+}, Mg^{2+}, Ne, Cl^-, S^{2-}, Ar, F^-, Na^+

4. Write three properties of metals that can help you distinguish them from non metals.

5. Circle the acidic oxides from the oxides given: Na_2O SO_2 NO_3 CaO

Ch 8/ Worksheet/Electron Configuration

Name: _____

Fill in these blanks about quantum numbers.

1. The maximum number of electrons in the third energy level (n=3) is _____.

2. Pair the following elements of electronic configurations which have the same properties.

Pairs	Electronic Configuration
	A. $1s^2\ 2s^2\ 2p^6\ 3s^2$
	B. $1s^2\ 2s^2\ 2p^3$
	C. $1s^2\ 2s^2\ 2p^6\ 3s^2\ 3p^6\ 4s^2\ 3d^{10}\ 4p^6$
	D. $1s^2\ 2s^2$
	E. $1s^2\ 2s^2\ 2p^6$
	F. $1s^2\ 2s^2\ 2p^6\ 3s^2\ 3p^3$

3. Name the elements whose electron configurations are:

Element	
	a. $1s^2\ 2s^2\ 2p^6\ 3s^2\ 3p^6$
	b. $1s^2\ 2s^2\ 2p^6\ 3s^2\ 3p^6\ 4s^2\ 3d^3$

4. Name the elements whose box electron configurations are:

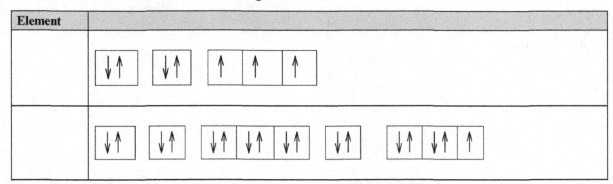

5. Without looking at the periodic table, write the spdf notation (electronic configuration) of the elements with the following atomic numbers:

   a. 10

   b. 22

   c. 28

6. Identify which element have the following electronic configurations?

   a. [Ar] $4s^1$ b. [Ar] $4s^2\ 3d^{10}\ 4p^3$

7. Write the spdf and box configuration for each of these. Do not use noble gas notation.
   a. potassium

   b. copper

   c. bromine

   d. manganese

8. Write the spdf and box configuration of the following ions. Do not use noble gas notation.
   a. Cr^{2+}

   b. I^-

Chapter 8: Electronic Configurations, Element Properties and the Periodic Table   251

9. Write and explain what is wrong with the following configurations? Write your answer clearly and give the correct answer are:

Element	
a. Al: $1s^2\ 2s^2\ 2p^6\ 3s^3\ 3p^1$	
b. V: $1s^2\ 2s^2\ 2p^6\ 3s^2\ 3p^6\ 3d^5$	
c. As: $1s^2\ 2s^2\ 2p^6\ 3s^2\ 3p^6\ 4s^2\ 4d^{10}\ 4p^3$	

10. What is wrong with the following configurations? Write your answer clearly and give the correct answer.

Element	
C Answer	[↓↑] [↑] [↑][↑][↑]
O Answer	[↓↑] [↓↑] [↓↑][↓↑][ ]
Ti Answer	[↓↑] [↓↑] [↓↑][↓↑][↓↑] [↓↑] [↓↑][↓↑][↓↑] [↑] [↑][↑][↑]
Br Answer	[Ar] [↓↑] [↓↑][↓↑][↓↑][↓↑][↓↑] [↑↑][↓↑][↑]

Ch 8/Worksheet/Periodic Properties   Name: _____

1. Choose four elements from rows 2 and 3 and write their isoelectric ionic structure.

2. Which of the following are isoeletric pairs?

   O^+, Ar, S^{2-}, Ne, Zn, Cs^+, N^{3-}, As^{3+}, N, Xe

3. Consider an element X. What will happen to the ionic radii for the following ions? Give a reasonable explanation to your answer.

   a. If it is X^{2+}

   b. If it is X^{3-}

4. Circle the larger of the ions in the following pairs:

   a. Cl or Cl^- b. Na or Na^+ c. Mg^{2+} or Al^{3+} d. Se^{2-} or Te^{2-}

5. Arrange in increasing order of atomic radii:

   a. Be, Mg, Ba   b. Al, Si, Ge   c. N^{3-}, O^{2-}, F^-

   d. Tl^{3+}, Tl^{2+}, Tl^+ e. K^+, P^{3-}, S^{2-}, Cl^-

6. Circle the element that has a higher ionization energy:

   Cl or Na Mg^{2+} or Mg   Cl or Br   P or Se

7. Identify the element that has the highest and lowest ionization energy?

   Rb, Mg, I, As, F

8. Identify the element that has the highest and lowest electron affinity.

   S, P, Ga, Li, Cl

9. Label the following as acidic or basic oxides.

   P_4O_{10} Fe_2O_3 Cr_2O_3

   MgO K_2O

Chapter 9
Chemical Bonds

· Lewis dot symbols · Octet rule

Ionic and covalent bonding

Ionic compounds

- Transfer of electrons. Metals give e- to form cations and non metals accept e- to form anions.
- Nomenclature of ionic compounds (metal and non-metal- ide)
- Ionic bonding also occurs in transition metals. (using roman numeral to designate the valency of transition metal in the name)

Covalent compounds

- Sharing of electrons.
- Lewis structures: single, double and triple bonds.
- Lone pair of electrons have to be shown.
- Nomenclature: using mono, di, tri etc to indicate number of atoms in molecule.
- Polar and non-polar covalent bonds
  - **Electronegativity** (EN): pull of electrons in a covalent bond.
  - Increases across the PT and decreases down PT.

Things to remember when writing Lewis structures of covalent compounds:

1. hydrogens and halogens are always terminal atoms
2. central atom usually has the lowest EN (or on the left of the PT)
3. in oxo-acids the hydrogens are usually bonded to oxygens
4. molecules and polyatomic molecules are usually compact and symmetrical.

Exceptions to octet rule

1. Odd number of electrons: when one e- is left unpaired: e.g. NO_2
2. Incomplete octets: when the atom does not have 8e- but is still stable: e.g. BF_3
3. Expanded valence shells: when atoms can have more than 8e-, for atoms that have the 3rd shell (n=3), e.g. SF_6

Bond lengths and bond energy: The more electrons are shared the tighter the atoms are held together hence short bond and high bond energy. E.g. bond length: single > double > triple; bond energy: triple > double > single.

Bond dissociation energy: energy required to break one mole of covalent bond in gas phase. Formation and breaking are opposite processes hence in one energy is released and in the other the same amount of energy is absorbed.

Resonance: movement of electrons in a bond. Delocalization, resonance hybrid

Formal Charge: (electronic book keeping) difference between the number of valence e in a free atom and number of e assigned to that atom when bonded.
FC= # of valence electrons in atom
 - # of lone pair electrons on the bound atom
 - ½ # of electrons in the bonds to the atom (or number of bonds on the atom) in a neutral compound/ molecule the FC should be zero.

Chapter 9 Bonding - 1

Dr. Sapna Gupta

Lewis Dot Symbol

- Lewis dot symbols is a notation where valence electrons are shown as dots.
- Draw the electrons symmetrically around the sides (top, bottom, left and right)

Notes:

Why Bonding? And Types?

- Bonding occurs to make elements more stable.
- Noble gases are stable, non reactive – it must be the electronic configuration.
- All elements try to have the noble gas configuration
- OCTET RULE: have 8 e- in the valence shell.

Bonding

Ionic
- Transfer of electrons
- Between metals and non metals
- Cations and Anions
- High melting points

Covalent
- Sharing of electrons
- Between non metals only
- No ions
- Low melting points

Notes:

Ionic Bond

- Made from transfer of electrons
- Metals lose electrons and become cations (metals have low ionization energy so lose electrons easily)
- Non metals gain electrons and become anions (non metals have high electron affinity so they gain electrons)
- Metals give electrons to non metals
- (Writing Lewis dot structures of ions)
- Below is a representation of transfer of electrons.

•Ca• + •Ö• → Ca^{2+} + $[:\ddot{O}:]^{2-}$

Lewis Structure – Ionic Compounds

- Sodium and Chlorine
  Na – needs to give one electron becomes Na^+
  Cl – needs to gain one electrons becomes Cl^-
  Na• + |Cl• Na^+ [|Cl|]$^-$
- Calcium and Fluorine
  Ca – needs to give two electron becomes Ca^{2+}
  F – needs to gain one electrons becomes F^-
  Ca: + |F• + |F• Ca^{2+} 2[|F|]$^-$
- Aluminum and Oxygen
  •Al• + |O $2Al^{3+}$ 3 [|O|]$^{2-}$

Covalent Compounds – Lewis Structures

- Atoms share electrons to form covalent bonds.

H• + •H ⟶ H:H or H–H

:C̈l• + :C̈l• ⟶ :C̈l:C̈l:

Notes:

Covalent Bonds

A **single bond** is a covalent bond in which one pair of electrons is shared by two atoms.

A **double bond** is a covalent bond in which two pairs of electrons are shared by two atoms.

A **triple bond** is a covalent bond in which three pairs of electrons are shared by two atoms.

Double bonds form primarily with C, N, and O.
Triple bonds form primarily with C and N.

:C̈l:C̈l: :C̈l – C̈l:
Ö::C::Ö   Ö=C=Ö
:N:::N: :N≡N:

Bond strength and bond length
bond strength   single < double < triple
bond length   single > double > triple

	N–N	N=N	N≡N
Bond Strength	163 kJ/mol	418 kJ/mol	941 kJ/mol
Bond Length	1.47 Å	1.24 Å	1.10 Å

Notes:

Writing Lewis Structures

1. Draw the skeleton structure of the molecule or ion by placing the lowest electronegative element in the center.
2. Add the total number of valence electrons. Subtract electron(s) if is a cation and add electron(s) if anion
3. Share one pair of electrons between each atom, and subtract those from the total number of electrons.
4. Distribute electrons to the atoms surrounding the central atom or atoms to satisfy the octet rule.
5. Distribute the remaining electrons as pairs to the central atom or atoms.
6. Add multiple bonds if atoms don't have the octet.

Hint:
H and halogens have single bonds (unless halogen is in the center)
O and S has two bonds (two single or one double)
N and P has three bonds (three single, one double one single or triple)
C has four bonds (different combination)

Steps for Writing Lewis Structure

Step	H_2O	CH_4	CO_2	O_2
1	H – O – H	H–C–H (with H above and below)	O – C – O	O – O
2	8	8	16	12
3	8-4 = 4	8-8 = 0	16 – 4 = 12	12 – 2 = 10
4	H – O – H	H–C–H (with H above and below)	:Ö – C – Ö:	:Ö – Ö:
5	H – Ö – H			
6			Ö = C = Ö	:Ö = Ö:

Exceptions to the Octet Rule

- Exceptions to the octet rule fall into three categories:
  - Molecules with an incomplete octet
  - Molecules with an expanded octet
  - Molecules with an odd number of electrons

Notes:

Incomplete Octets

Example: BF_3 (boron trifluoride)

$$BF_3 \Rightarrow (1 \times 3) + (3 \times 7) = 24 \text{ val. } e^-$$

$:\!\ddot{F} - \underset{|}{\overset{:\ddot{F}:}{B}} - \ddot{F}\!:$ (no octet) $:\!\ddot{F} - \underset{-1}{\overset{:\ddot{F}:}{B}} = \ddot{F}\!:^{+1}$

- Common with Be, B and Al compounds, but they often dimerize or polymerize.

Example: Be–Cl–Be–Cl–Be–Cl–Be (with bridging Cl atoms above and below)

Notes:

Expanded Octet

Elements of the 3rd period and beyond have *d*-orbitals that allow more than 8 valence electrons.

SF_6 = [Lewis structure with S central atom bonded to 6 F atoms]   48 valence e^-
(S has 12 valence electrons)

XeF_2 = :F̈–Xë–F̈:   22 valence e^-
(Xe has 10 valence electrons)

Odd Numbers of Electrons

Example: NO (nitrogen monoxide or nitric oxide)

NO ⇒ (1 × 5) + (1 × 6) = 11 valence e^-

:N=Ö: ↔ :N=Ö: Are these both equally good?
  better

Example: NO_2 (nitrogen dioxide)

NO_2 ⇒ (1 × 5) + (2 × 6) = 17 val. e^-

:Ö=N–Ö: ↔ :Ö–N=Ö: ↔ :Ö=N–Ö: ↔ :Ö–N=Ö:
   best
          Are these all equally good?

Chapter 9: Chemical Bonds

Key Points

- Lewis dot symbols
- Ionic bonding
- Covalent bonding
- Octet rule
- Lewis structures
- Exceptions to the Octet Rule
  - Incomplete octets
  - Expanded octets
  - Odd numbers of electrons

Notes:

Ch 9/ PowerPoint Study-1 Bonding - Lewis Structures   Name: _____

Answer these questions as you are watching the videos. They are due in class.
These questions are not just for you to answer but also to prepare you for the exam.
Make sure you understand what you are writing and not just copy from the text book. **Show all work.**

1. Identify the following elements as metals or nonmetals and write their Lewis structures in the row below.

Element	Li	F	S	Ca	N
Metal or Non metal					
Lewis Structure					

2. Write the Lewis structures of the following ions.

Li^+	S^{2-}	Ca^{2+}	N^{3-}	F^-

Use the strategies given below for writing Lewis structures.
a. First determine if you are writing Lewis structure for ionic or covalent compound. If ionic, i.e. metal and nonmetal, then see how to draw the Lewis structure from the power point. For covalent (all nonmetals), see steps 2 onwards.
b. Identify the central atom in the covalent compound – this is usually the left one in the compound or left in the periodic table.
c. Write the atom and write all the valence electrons around it.
d. Place all the other atoms surrounding the central atom. These are also called the terminal atoms.
e. Write all the valence electrons on the terminal atoms.
f. See how many electrons need to be shared for the terminal atoms with the central atom to give the terminal atom an octet (or duet in case of H).

3. From the elements in question 1, pick the element pairs that will form ionic and the ones that will form covalent bonds (table continued on the next page); write the compound formed and the Lewis structure.

Ionic Bond Pairs		Lewis Structure	
Pair: Select a metal and nonmetal	Compound formed		

Covalent Bond Pairs		Lewis Structure	
Pair	Compound formed		
N and F	NF_3		
F and S	SF_2		

263

Chapter 9: Chemical Bonds

4. For the following compounds, count the total number of electrons on the <u>central atom</u> and write which octet rule is being violated: incomplete shell, expanded shell, or unpaired electron. If no rule is violated, then write so.

   Strategy: a) Count the number of bonds and multiply by 2, b) add any lone pair electrons left on the atom. Use the total number of electrons to find the violations.

Compound	
H–P(H)(H)(H)(H) (5 H bonded to P)	
Cl–B(Cl)(Cl)	
HO–N=O (with lone pair on N)	

Ch 9/ PowerPoint Study-2 Lewis Structures

Name: _____

Answer these questions as you are watching the videos. They are due in class.
These questions are not just for you to answer but also to prepare you for the exam.
Make sure you understand what you are writing and not just copy from the text book. **Show all work.**

Draw the Lewis structure for the following compounds. Use the strategy given here to assist in your work.

1. First determine if you are writing Lewis structure for ionic or covalent compound. If ionic, i.e. metal and nonmetal, then see how to draw the Lewis structure from the power point. For covalent (all nonmetals), see steps 2 onwards.
2. Identify the central atom – this is usually the left one in the compound or left in the periodic table.
3. Write the atom and write all the valence electrons around it.
4. Place all the other atoms surrounding the central atom. These are also called the terminal atoms.
5. Write all the valence electrons on the terminal atoms.
6. See how many electrons need to be shared for the terminal atoms with the central atom to give the terminal atom an octet (or duet in case of H).

BI_3 AsF_5 ClO_2

OCS KI

(Write OCS using C as the central atom)

Challenge compounds – do these once you are feeling comfortable with the compounds above.

CH_3COOH	$HClO_3$
There are 3 central atoms: C, C and O All atoms are connected as listed except the first oxygen – it is bonded only to the central C.	Cl is the central atom. Place all other atoms around it. Make sure all oxygens are bonded to Cl and the H is bonded to one of the oxygens. Remember: Cl is capable of having an expanded shell.

Chapter 9 Bonding – 2 Polar Covalent Bond, Electronegativity, Formal Charge, Resonance

Dr. Sapna Gupta

Polar Covalent Bond

- **Nonpolar covalent bond** = electrons are shared *equally* by two bonded atoms
- **Polar covalent bond** = electrons are shared *unequally* by two bonded atoms.

M:X	$M^{\delta+}X^{\delta-}$	M^+X^-
Covalent bond	Polar covalent bond	Ionic bond
Equal sharing of electrons of the same electronegativity	Partially charged atoms held by unequally sharing electrons	Cations and anions held by electrostatic forces

$\delta+ \quad \delta-$
$H - F$

$\overset{\longrightarrow}{H - F}$

alternate representations

Electronegativity

- **Electronegativity**, X, is a measure of the ability of an atom in a molecule to attract bonding electrons to itself. Electronegativity increases across the group and decreases down a group.

	IA	IIA											IIIA	IVA	VA	VIA	VIIA
	1 H 2.1				Increasing Electronegativity →												
	3 Li 1.0	4 Be 1.5											5 B 2.0	6 C 2.5	7 N 3.0	8 O 3.5	9 F 4.0
	11 Na 0.9	12 Mg 1.2	IIIB	IVB	VB	VIB	VIIB		VIIIB		IB	IIB	13 Al 1.5	14 Si 1.8	15 P 2.1	16 S 2.5	17 Cl 3.0
	19 K 0.8	20 Ca 1.0	21 Sc 1.3	22 Ti 1.5	23 V 1.6	24 Cr 1.6	25 Mn 1.5	26 Fe 1.8	27 Co 1.9	28 Ni 1.9	29 Cu 1.9	30 Zn 1.6	31 Ga 1.6	32 Ge 1.8	33 As 2.0	34 Se 2.4	35 Br 2.8
	37 Rb 0.8	38 Sr 1.0	39 Y	40 Zr	41 Nb	42 Mo	43 Tc	44 Ru	45 Rh	46 Pd	47 Ag	48 Cd	49 In 1.7	50 Sn 1.8	51 Sb 1.9	52 Te 2.1	53 I 2.1
	55 Cs 0.7	56 Ba 0.9	57 La	72 Hf	73 Ta	74 W	75 Re	76 Os	77 Ir	78 Pt	79 Au 1.9	80 Hg 1.8	81 Tl 1.9	82 Pb	83 Bi	84 Po 2.0	85 At 2.2

Decreasing Electronegativity ↓

Source: Sapna Gupta

Polar Bond, Contd...

The difference in electronegativity between the two atoms in a bond is a rough measure of bond polarity.

When the difference is very large, an ionic bond forms. When the difference is large, the bond is polar. When the difference is small, the bond is nonpolar.

Solved Problem:
Using electronegativities, arrange the following bonds in order by increasing polarity: C—N, Na—F, O—H.

For C—N, the difference is 3.0 (N) − 2.5 (C) = 0.5.
For Na—F, the difference is 4.0 (F) − 0.9 (Na) = 3.1.
For O—H, the difference is 3.5 (O) − 2.1 (H) = 1.4.
Bond Polarities: C-N < O-H < Na-F

Solved Problem:
What is the Lewis structure of NO_3^- ?

1) Draw skeletal structure with central atom being the least electronegative.

    $$O - \overset{O}{\underset{|}{N}} - O$$

2) Add valence electrons. Add 1 for each negative charge and subtract 1 for each positive charge.

    $NO_3^- \Rightarrow (1 \times 5) + (3 \times 6) + 1 = 24$ valence e^- 24 e^-

3) Subtract 2 for each bond in the skeletal structure.   − 6 e^-

4) Complete electron octets for atoms bonded to the central atom except for hydrogen.

    $$:\ddot{O} - \overset{..}{\underset{|}{N}} - \ddot{O}:$$ 18 e^-

5) Place extra electrons on the central atom.

6) Add multiple bonds if atoms lack an octet.

    $$\left[:\ddot{O} - \overset{:\ddot{O}:}{\underset{|}{N}} = \ddot{O}: \right]^-$$ 24 e^-

Lewis Structures and Formal Charge

The electron surplus or deficit, relative to the free atom, that is assigned to an atom in a Lewis structure.

$$\text{Formal Charge} = \text{Total valence electrons} - \text{Total non-bonding electrons} - \frac{1}{2}\left(\text{Total bonding electrons}\right)$$

Example: H_2O = H:Ö:H

H: orig. valence e^- = 1	O: orig. valence e^- = 6
− non-bonding e^- = −0	− non-bonding e^- = −4
− 1/2 bonding e^- = −1	− 1/2 bonding e^- = −2
formal charge = 0	formal charge = 0

Formal charges are not "real" charges.

Formal Charge, contd..

- A Lewis structure with *no* formal charges is generally better than one with formal charges.
- Small formal charges are generally better than large formal charges.
- Negative formal charges should be on the more electronegative atom(s).

Example: H_2CO

H C O H   or   H C O H   ?

Answer:

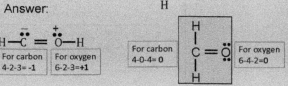

For carbon 4-2-3= -1   For oxygen 6-2-3=+1   For carbon 4-0-4= 0   For oxygen 6-4-2=0

Chapter 9: Chemical Bonds

Resonance

Resonance is the movement of electrons over two or more bonds. In these cases two or more equally valid Lewis structures can be written.

Example: NO_2^-

$$[:\ddot{O} - \ddot{N} = \ddot{O}:]^-$$

These two bonds are known to be identical.

Solution: $[:\ddot{O} - \ddot{N} = \ddot{O}:]^- \longleftrightarrow [:\ddot{O} = \ddot{N} - \ddot{O}:]^-$

Two **resonance structures**, their average or the **resonance hybrid**, best describes the nitrite ion.
The double-headed arrow indicates resonance.

Notes:

Key Points

- Lewis structures
- Bond polarity
- Electronegativity
- Formal charge
- Resonance structures

Notes:

Ch 9/ PowerPoint Study-3 Formal Charge/Polarity   Name: _____

Answer these questions as you are watching the videos. They are due in class.
These questions are not just for you to answer but also to prepare you for the exam.
Make sure you understand what you are writing and not just copy from the text book. **Show all work.**

1. What is meant by electronegativity?

2. Circle the element in the pairs below with higher electronegativity.

    Li or Be   N or O   P or N   I or F

3. Which of the following bonds will be polar covalent? (If you want to take the next step, write δ⁺ and δ⁻ on the elements. Remember –δ⁻ is on the element of higher electronegativity)

    N-F   O=O   N-O

4. Calculate the formal charges on all the atoms of the ions given below. Use the strategy given:
    a. Write the Lewis structure of the compounds/ions.
    b. Make sure you have the right number of electrons on each atom – all the bonding and non bonding. (Use the worksheet from class to make sure you can draw the structures).
    c. Use the formula given below to calculate the formal charge on each atom.

    Valence electrons – non bonding electrons – number of bonds on the atom = formal charge
    (Formal charge will range from -1 to +1)

1) Carbonate Ion	2) ICl_4^-
3) PS_3^{-1}	4) P_2H_4

Ch 9/Worksheet/Bonding-Lewis Structures   Name: _____

1. Identify the following elements as metals or nonmetals and write their Lewis structures in the row below.

Element	Al	Cl	O	Mg	Cs	P
Metal or Non metal						
Lewis Structure						

2. Write the Lewis structures of the following ions.

Li^+	O^{2-}	Mg^{2+}	Al^{3+}	N^{3-}	Cl^-

3. From the elements in question 1, pick the element pairs that will form ionic and the ones that will form covalent bonds; write the compound formed and the Lewis structure.

Ionic Bond Pairs		Lewis Structure
Pair	Compound formed	

Covalent Bond Pairs		Lewis Structure
Pair	Compound formed	

4. Identify what is wrong with the following Lewis structures and give the correct structure.

Structure	Explain what is wrong	Correct structure
H–C̈=N̈		
:F̈–B(–F̈:)–F̈: (trigonal with three F around B)		
:F̈–N(–F̈:)–F̈: (trigonal with three F around N)		
·H=C=C=H		
H–Ö=F̈:		
Ö–Sn–Ö		
H, :Ö : attached to C–F̈:		

Ch 9/Worksheet/Bonding – Lewis Structures

Name: _____

1. For each of the following molecules, draw the Lewis structure

1) carbon tetrafluoride	2) BF_3
3) NF_3	4) H_2CS
5) O_2	6) CH_2F_2
9) PF_3	10) H_2S
11) CF_2S	12) SHF

Chapter 9: Chemical Bonds

2. Writing Lewis structures for ions and compounds with multiple central atoms.
   a. Write the central atom.
   b. Write all the other atoms around it (terminal atoms)
   c. Bond them all with at least one bond.
   d. Follow the rule given in the power point to determine how many bonds are needed for each atom (e.g. H and halogens, 1 bond; O is 2 bonds)
   e. Assign all the valence electrons to all the atoms as needed.
   f. Assign any remaining electrons to oxygen or the more electronegative atom.

   As a failsafe, do these steps: 1) Add ALL the valence electrons of ALL the atoms. 2) If there is ONE negative charge, add one electron; if there is a ONE positive charge, subtract one electron. 3) After your final structure is drawn, add ALL the electrons in the compound. They should be the same as in step 1.

1) Sulfate Ion	2) Nitrate Ion
3) $CH_3CH_2NH_3^+$	4) ClO_4^-

Ch 9/Worksheet/Formal Charges and Resonance Structures

Name: _____

Calculating Formal Charge

Formal Charge = Number of valence e^- – (1/2 number of bonding e^- + lone pair of e^-)

OR Formal Charge = Number of valence e^- – (number of bonds + lone pair of e^-)

Resonance Structures: Determining if a structure is more stable or possible than the other

Resonance structures are structures where electrons can move around in a molecule. Electrons move only in double or triple bonds; electrons can also move in lone pairs. No bonds can be broken. No atoms can move around.

1. Calculate the formal charges on each atom;
2. If the molecule is electrically neutral, then structure is stable.
3. If more electronegative element has the negative charge then structure is more stable.
4. There should not be a positive and a negative next to each other – that destabilizes the structure.

---

1. Draw the Lewis structures of the following ions:

   SO_4^{2-} CN^- $CH_3NH_3^+$

2. What are the formal charges on all the atoms following structures?

   :Ö-Cl(=O)-Ö: :Ö–Ö=Ö: :C≡O:

3. Which of the following are resonance pairs in the left column? If they are then which is more likely to occur? Why?

$[:N \equiv C - \ddot{O}:]^-$ or $[:\ddot{N} = C = \ddot{O}:]^-$	
H–C≡N: :C≡N–H	
(acetone with C=O) H₃C–C(=O)–CH₃ (acetone with C⁺–O⁻) H₃C–C⁺(–O:)–CH₃	

277

Chapter 10
Chemical Bonding II: Molecular Geometry and Bonding Theory

Ionic bonding: crystal lattice
Covalent bonding: VSEPR, valence bond and hybridization theory
Metallic bonding: ability of electrons to flow on all atoms.

VSEPR: Valence shell electron pair repulsion theory

Electrons try to be as far away from each other as possible. In bonding and bonding theory only the valence electrons are important. Each geometry is associated with its bond angles.
Molecular geometry can be as follows:
(A = central atom, X = terminal atoms and E = lone pair of electrons)

Electron Groups	AXE formula	Bond Angle	Example	Electronic Geometry	Shape of Molecule
2	AX_2	180°	$BeCl_2$	Linear	Linear
3	AX_3	120°	BF_3	Trigonal planar	Trigonal planar
3	AX_2E	120°	SO_2	Trigonal planar	Bent
4	AX_4	109.5°	CH_4	Tetrahedral	Tetrahedral
4	AX_3E	109.5°	NH_3	Tetrahedral	Trigonal Pyramidal
4	AX_2E_2	109.5°	H_2O	Tetrahedral	Bent
5	AX_5	90°, 120°, 180°	PCl_5	Trigonal bipyramidal	Trigonal Bipyramidal
5	AX_4E	90°, 120°, 180°	SF_4	Trigonal bipyramidal	Seesaw
5	AX_3E_2	90°, 180°	ClF_4	Trigonal bipyramidal	T – shape
5	AX_2E_3	180°	XeF_2	Trigonal bipyramidal	Linear
6	AX_6	90°, 180°	SF_6	Octahedral	Octahedral
6	AX_5E	90°	BrF_5	Octahedral	Square Pyramidal
6	AX_4E_2	90°	XeF_4	Octahedral	Square Planar
6	AX_3E_3	90°, 180°		Octahedral	T – Shape
6	AX_2E_4	180°		Octahedral	Linear

Figure on the next page.

Chapter 10: Chemical Bonding II: Molecular Geometry and Bonding Theory

No of e- groups	Geometry (all atoms)	1 Lone Pair	2 Lone Pairs	3 Lone Pairs	4 Lone Pairs
2	Linear AX_2				
3	Trigonal Planar AX_3	Bent AX_2E			
4	Tetrahedral AX_4	Trigonal pyramidal AX_3E	Bent AX_2E_2		
5	Trigonal Bipyramidal AX_5	Seesaw AX_4E	T-Shape AX_3E_2	Linear AX_2E_3	
6	Octahedral AX_6	Square Pyramid AX_5E	Square Planar AX_4E_2	T-shape AX_3E_3	Linear AX_2E_4

Source: Sapna Gupta

Dipole moment and polar molecules

- Polar bond and polar molecules
- Polar bond – when the two atoms of a bond have different electronegativity while
- Polar molecules have overall non-zero dipole moments and non polar molecules have zero dipole moment.
- A bond can be polar but the molecule does not have to be.
- Dipole moment = magnitude of charge x distance separating the positive and negative charge
- Units (debye = coulomb x meter)
- Dipole moment usually can be determined by exposing molecules to electric field.

Chapter 10: Chemical Bonding II: Molecular Geometry and Bonding Theory

Determining the polarity of a molecule
1. write Lewis structure of molecule (write ALL electrons)
2. write the shape of molecule
3. check the electronegativity of each atom (estimate from PT)
4. see the symmetry of the molecule (not just in shape but in types of atoms)
5. see if the equal electronegativity of similar atoms cancel each other out, if not then molecule may be polar.

Valence Bond Theory

Electron clouds of opposite spins overlap to form bonds. Important points:
a. Electrons stay in their respective orbitals
b. Bonding electrons localize in the region of overlap
c. Maximum overlap occurs when orbitals overlap end to end
d. Molecular geometry depends on geometry of orbitals.

Has flaws when looking at bond angles and molecular of majority of molecules.

Hybridization

Mixing of orbitals of excited atoms to form new orbitals, which overlap to form bonds.

Type	Typical Shape	Bond Angle	e.g.
Sp^3	Tetrahedral	109°	CH_4, HN_3
Sp^2	Trigonal planar	120°	BF_3, $CH_2=CH_2$
Sp	Linear	180°	BeF_2, $CH\equiv CH$

Hybrid orbitals of d shell are also possible.
Hybridization in organic chemistry
Alkanes, alkenes, alkynes.
Geometric isomerism: cis and trans geometry of alkenes.

Molecular Orbital Theory

Atomic orbitals (AO) combine to form molecular orbitals (MO) and antibonding molecular orbitals.

Chapter 10 Shapes of Molecules

Dr. Sapna Gupta

Shapes of Molecules - Importance

- All molecules have a 3D orientations; even the diatomic ones because atoms have a volume.
- In case of tri atomic or polyatomic molecules and ions these shapes can get very important.
- Physical properties of molecules can be predicted by the shape of molecule.
    - Why is H_2O liquid but CO_2 a gas at room temperature?
- Molecular interactions can be predicted by shape of molecule.
- A number of biological functions occur because of proper molecular interactions.
    - Hemoglobin and oxygen binding

Notes:

Hemoglobin

- 4 Protein subunits + 4 macromolecules (with metal) = hemoglobin
- Macromolecule is porphyrin with iron in the center for bonding with oxygen.

Sickle cell anemia – inability of hemoglobin to bind to oxygen. Difference of one amino acid (glutamic acid is replaced by valine) changes the shape of the whole protein.

(see https://evolution.berkeley.edu/evolibrary/article/mutations_06 for structures)

Notes:

Chapter 10: Chemical Bonding II: Molecular Geometry and Bonding Theory

Molecular Geometry

- This is the three dimensional shape of a molecule.
- Geometry can be predicted by Lewis structures and VSEPR theory.
- VSEPR – Valence Shell Electron Pair Repulsion Theory. This theory indicates that electron pairs, bonding or non bonding on the central atom, move far away to minimize repulsion.
- Predict the geometry using the strategy below.

Lewis structure → Electron-domain geometry → Molecular geometry

Notes:

VSEPR- Electron Geometry 1

- Two electron pairs are 180° apart (a linear arrangement).
- Three electron pairs are 120° apart in one plane (a trigonal planar arrangement).
- Four electron pairs are 109.5° apart in three dimensions (a tetrahedral arrangement).

Number of pairs of electrons	2	3	4
Arrangement	Linear	Trigonal Planar	Tetrahedral

© magnetix/Shutterstock.com

Notes:

Chapter 10: Chemical Bonding II: Molecular Geometry and Bonding Theory

VSEPR – Electron Geometry 2

- Five electron pairs are arranged with three pairs in a plane 120° apart and two pairs at 90° to the plane and 180° to each other (a trigonal bipyramidal arrangement).
- Six electron pairs are 90° and 180° to each other apart (an octahedral arrangement).

Number of pairs of electrons	5	6
Arrangement	Trigonal Bipyramidal	Octahedral

Notes:

VSEPR – Electronic Geometry 3

Space Filled Model	∞	♣	✦	✦	✦
Shape	Linear	Trigonal Planar	Tetrahedral	Trigonal Bipyramidal	Octahedral
Number of e⁻ groups / AXE formula	2 e⁻ groups AX_2	3 e⁻ groups AX_3	4 e⁻ groups AX_4	5 e⁻ groups AX_5	6 e⁻ groups AX_6
Example	BeH_2	SO_3	CH_4	IF_5	SF_6

NOTE: AXE formula is used to represent the number of electron groups around the central atom. "**A**" is the central atom, "**X**" is the number of atoms attached and "**E**" is number of electron pairs on the central atom.

Source: Sapna Gupta

Notes:

Chapter 10: Chemical Bonding II: Molecular Geometry and Bonding Theory

Shapes Possible – For AX_2 and AX_3

	Electronic geometry	Bond angle	Shape	Model
AX_2	Linear	180°	Linear	
AX_3	Trigonal planar	120°	Trigonal planar	
AX_2E	Trigonal planar	120°	Bent (The green balloon is the lone pair of electron)	

Notes:

Shapes Possible – AX_4

	Electronic geometry	Bond angle	Shape	Model
AX_4	Tetrahedral	109°	Tetrahedral	
AX_3E	Tetrahedral	109°	Trigonal pyramidal (The blue balloon is the lone pair of electron)	
AX_2E_2	Tetrahedral	109°	Bent (The blue balloon is the lone pair of electron)	

Notes:

Chapter 10: Chemical Bonding II: Molecular Geometry and Bonding Theory

Shapes Possible – AX_5 and AX_6

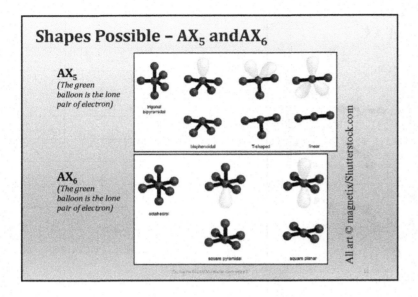

AX_5 (The green balloon is the lone pair of electron): trigonal bipyramidal, bisphenoidal, T-shaped, linear

AX_6 (The green balloon is the lone pair of electron): octahedral, square pyramidal, square planar

Electron Groups	AXE formula	Bond Angle	E.g.	Electronic Geometry	Shape of Molecule
2	AX_2	180°	$BeCl_2$	Linear	Linear
3	AX_3	120°	BF_3	Trigonal planar	Trigonal planar
3	AX_2E	120°	SO_2	Trigonal planar	Bent
4	AX_4	109.5°	CH_4	Tetrahedral	Tetrahedral
4	AX_3E	109.5°	NH_3	Tetrahedral	Trigonal Pyramidal
4	AX_2E_2	109.5°	H_2O	Tetrahedral	Bent
5	AX_5	90°, 120°, 180°	PCl_5	Trigonal bipyramidal	Trigonal Bipyramidal
5	AX_4E	90°, 120°, 180°	SF_4	Trigonal bipyramidal	Seesaw
5	AX_3E_2	90°, 180°	ClF_3	Trigonal bipyramidal	T – shape
5	AX_2E_3	180°	XeF_2	Trigonal bipyramidal	Linear
6	AX_6	90°, 180°	SF_6	Octahedral	Octahedral
6	AX_5E	90°	BrF_5	Octahedral	Square Pyramidal
6	AX_4E_2	90°	XeF_4	Octahedral	Square Planar
6	AX_3E_3	90°, 180°		Octahedral	T – Shape
6	AX_2E_4	180°		Octahedral	Linear

Chapter 10: Chemical Bonding II: Molecular Geometry and Bonding Theory

VSEPR Geometries

No of e-groups	Geometry (all atoms)	1 Lone Pair	2 Lone Pairs	3 Lone Pairs	4 Lone Pairs
2	Linear AX_2				
3	Trigonal Planar AX_3	Bent AX_2E			
4	Tetrahedral AX_4	Trigonal pyramidal AX_3E	Bent AX_2E_2		
5	Trigonal Bipyramidal AX_5	Seesaw AX_4E	T-Shape AX_3E_2	Linear AX_2E_3	
6	Octahedral AX_6	Square Pyramid AX_5E	Square Planar AX_4E_2	T-shape AX_3E_3	Linear AX_2E_4

Source: Sapna Gupta

Notes:

Solved Problem:
Use the VSEPR model to predict the geometries of the following molecules:
a. AsF_3
b. PH_4^+

AsF_3 has $1(5) + 3(7) = 26$ valence electrons; As is the central atom.

There are 4 pairs of electrons around As; three bonding and one lone pair.
AX_3E
The electronic geometry is tetrahedral. One of these regions is a lone pair, so the molecular geometry is trigonal pyramidal.

PH_4^+ has $1(5) + 4(1) - 1 = 8$ valence electrons; P is the central atom.

There are 4 pairs of electrons around P; all four bonding electron pairs.
AX_4
The electronic geometry is tetrahedral. All regions are bonding, so the molecular geometry is tetrahedral.

Notes:

Chapter 10: Chemical Bonding II: Molecular Geometry and Bonding Theory

Solved Problem:
Use the VSEPR model to predict the geometries of the following molecules:
a. ICl_3
b. ICl_4^-

ICl_3 has $1(7) + 3(7) = 28$ valence electrons. I is the central atom.

There are five regions: three bonding and two lone pairs.
AX_3E_2
The electronic geometry is trigonal bipyramidal.
The geometry is T-shaped.

ICl_4^- has $1(7) + 4(7) + 1 = 36$ valence electrons. I is the central element

There are six regions around I: four bonding and two lone pairs.
AX_4E_2
The electronic geometry is octahedral.
The geometry is square planar.

Notes:

Molecule with more than one Central Atom

For CH_3OH there are two central atoms so each will have its own geometry.

CH_3OH

No lone pairs AX_4 AX_4 2 lone pairs
tetrahedral   bent

Notes:

Chapter 10: Chemical Bonding II: Molecular Geometry and Bonding Theory

Dipole Moment and Polarity of Molecule

- Polarity is a degree of charge separation in a molecule
- For HCl, we can represent the charge separation using δ+ and δ- to indicate partial charges. Because Cl is more electronegative than H, it has the δ- charge, while H has the δ+ charge.

$$\overset{\delta+}{H}\text{——}\overset{\delta-}{Cl}$$

- Dipole moment a measure of how much a molecule can move in an electrical field. The movement occurs only if there is a charge separation.
- Polar molecules have dipole moment, while non polar molecules have zero dipole moment.
- To determine dipole moment:
    1. Draw the Lewis structure
    2. Determine the molecular shape of the molecule
    3. Determine the electronegativity from the periodic table
    4. See if the molecule is symmetrical as that will nullify the charge separation.
    5. Determine if the molecule is polar of not (yes – if molecule is asymmetric)

Notes:

- **Molecules with more than two atoms**
    - Remember bond dipoles are additive since they are *vectors*.

H_2O — dipole moment > 0

BeH_2 — dipole moment = 0

Example: Dichloroethene, $C_2H_2Cl_2$, exists as three isomers.

cis-1,2-dichloroethene	trans-1,2-dichloroethene	1,1-dichloroethene
polar	nonpolar	polar
μ = 1.90 D	μ = 0 D	μ = 1.34 D
bp = 60.3 °C	bp = 47.5 °C	bp = 31.7 °C

Notes:

Top-row art © magnetix/Shutterstock.com

Chapter 10: Chemical Bonding II: Molecular Geometry and Bonding Theory

Solved Problem:
Which of the following molecules have dipole moment?
a. GeF_4 b. SF_2 c. AsF_3

GeF_4: $1(4) + 4(7) = 32$ valence electrons.
Ge is the central atom.
8 electrons are bonding; 24 are nonbonding.
Tetrahedral molecular geometry. (AX_4)

GeF_4 is nonpolar and has no dipole moment.

SF_2: $1(6) + 2(7) = 20$ valence electrons.
S is the central atom.
4 electrons are bonding; 16 are nonbonding.
Bent molecular geometry. (AX_2E_2)

SF_2 is polar and has a dipole moment.

AsF_3: $1(5) + 3(7) = 26$ valence electrons.
As is the central atom.
6 electrons are bonding; 20 are nonbonding.
Trigonal pyramidal molecular geometry. (AX_3E)

AsF_3 is polar and has a dipole moment.

Key Words/Concepts

- Molecular Geometry
- Shapes/VSEPR
- AXE formula
- Bonding and non bonding electrons
- Bond angles
- Electronegativity
- Bond polarity
- Polarity of molecule
- Dipole moment

Ch 10/ PowerPoint Study-1-Molecular Geometry   Name: _____

Answer these questions as you are watching the videos. They are due in class.
These questions are not just for you to answer but also to prepare you for the exam.
Make sure you understand what you are writing and not just copy from the text book. **Show all work.**
In each case, predict (a) the shape of the molecule and b) the bond angle. Follow the strategy:

1. Write the Lewis structure – make sure you show ALL the electrons.

2. Count the number of electron "groups" on the central atom (underlined) – a single bond, double bond, triple bond and one lone pair are each one group. So four single bonds are four electron groups; three single bonds and one lone pair are also four electron groups – see the power point for all options.

3. From the power point table, identify which "geometry" the central atom follows.

4. Write the AXE formula etc as asked in the table below.

5. Do g) after you finish learning about polarity of molecules.

   Do all three, step by step.

Molecule→	(1) $\underline{O}F_2$	(2) $H_2\underline{C}O$	(4) $\underline{B}F_3$
a) Lewis structure			
b) Electron groups on central atom.			
c) AXE formula			
d) Electronic geometry on the central atom			
e) Bond angle associated with the geometry			
f) Shape of the molecule			
g) Polar (yes/no)			

Chapter 10 Bonding Theories

Dr. Sapna Gupta

Valence Bond Theory

- Valence bond theory is an approximate theory put forth to explain the covalent bond by quantum mechanics.

A bond forms when
- An orbital on one atom comes to occupy a portion of the same region of space as an orbital on the other atom. The two orbitals are said to overlap.
- The total number of electrons in both orbitals is no more than two. The greater the orbital overlap, the stronger the bond.
- Orbitals (except s orbitals) bond in the direction in which they protrude or point, so as to obtain maximum overlap.

Notes:

Methane Molecule According to Lewis Theory

- CH_4
- C is central atom: $1s^2\ 2s^2\ 2p^2$
- Valence electrons for bonding are in the s (spherical) and p (dumbbell) orbitals.
- The orbital overlap required for bonding will be different for the two bonds.
- Two bonds will be longer and two shorter and the bond energy will be different too.
- Therefore there must be a different theory on how covalent bonds are formed.

Source: Sapna Gupta

Notes:

Chapter 10: Chemical Bonding II: Molecular Geometry and Bonding Theory

Valence Bond Theory - Hybridization

Hybrid orbitals are formed by mixing orbitals, and are named by using the atomic orbitals that combined:
- one *s* orbital + one *p* orbital gives two *sp* orbitals
- one *s* orbital + two *p* orbitals gives three sp^2 orbitals
- one *s* orbital + three *p* orbitals gives four sp^3 orbitals
- one *s* orbital + three *p* orbitals + one *d* orbital gives five sp^3d orbitals
- one *s* orbital + three *p* orbitals + two *d* orbitals gives six sp^3d^2 orbitals

Notes:

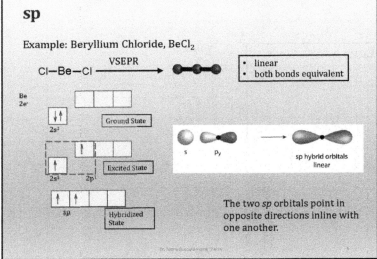

Notes:

All art © magnetix/Shutterstock.com

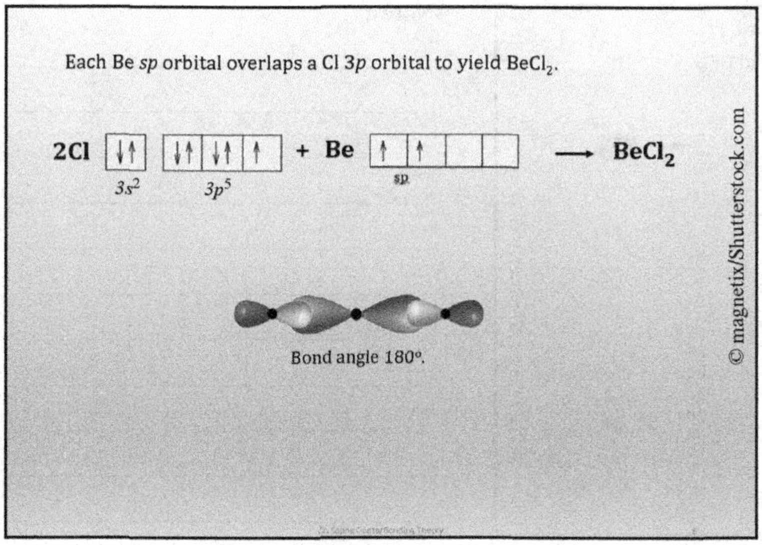

Chapter 10: Chemical Bonding II: Molecular Geometry and Bonding Theory   295

Each Be *sp* orbital overlaps a Cl 3*p* orbital to yield $BeCl_2$.

2Cl [↑↓] [↑↓][↑↓][↑] + Be [↑][↑][ ][ ] ⟶ $BeCl_2$
    $3s^2$ $3p^5$ sp

Bond angle 180°.

Notes:

---

sp^2

Example: Boron trifluoride, BF_3

:F:
 |
 B   → VSEPR → • trigonal planar
/ \   • all bonds equivalent
:F: :F:

B
3e⁻

[↑↓] [↑][ ][ ]   Ground State
$2s^2$ $2p^1$

[↑]   [↑][↑][ ]   Excited State
$2s^1$ $2p^2$

[↑][↑][↑][ ]   Hybridized State
  sp^2

The three sp^2 orbitals point to the corners of an equilateral triangle.

s + p_x, p_y → sp^2 hybrid orbitals trigonal planar

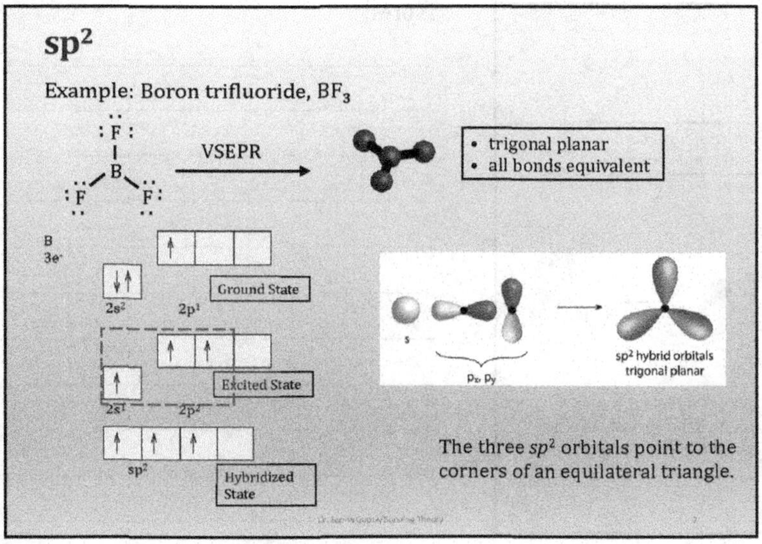

Notes:

All art © magnetix/Shutterstock.com

Chapter 10: Chemical Bonding II: Molecular Geometry and Bonding Theory

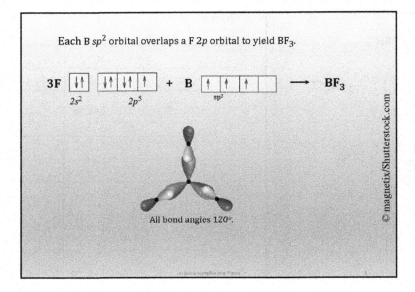

Notes:

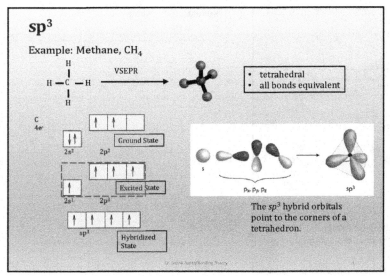

Notes:

All art © magnetix/Shutterstock.com

Chapter 10: Chemical Bonding II: Molecular Geometry and Bonding Theory

Each C $2sp^3$ orbital overlaps a H $1s$ orbital to yield CH_4.

4 H [↑] $1s^1$ + C [↑][↑][↑][↑] sp^3 ⟶ CH_4

All bond angles 109.5°.

Working out Hybridization

To figure out hybridization on the central atom in a molecule:
1. Write the Lewis electron-dot formula.
2. Use VSEPR to determine the electron geometry about the atom.
3. From the electronic geometry deduce the hybrid orbitals.
4. Assign the valence electrons to the hybrid orbitals one at a time, pairing only when necessary.
5. Form bonds by overlapping singly occupied hybrid orbitals with singly occupied orbitals of another atom.

Chapter 10: Chemical Bonding II: Molecular Geometry and Bonding Theory

Determining Hybridization

Lewis structure → Number of electron domains → Type of hybridization

Electron groups	Hybrid orbitals	Electronic Geometry	3D shape
2	Sp	Linear	
3	Sp²	Trigonal planar	
4	Sp³	Tetrahedral	
5	Sp³d	Trigonal bipyramidal	
6	Sp³d²	Octahedral	

Notes:

Hybridized Orbital Characteristics

Electron Groups	AXE formula	Bond Angle	E.g.	Electronic Geometry	Hybridization	Shape of Molecule
2	AX_2	180°	$BeCl_2$	Linear	sp	Linear
3	AX_3	120°	BF_3	Trigonal planar	sp²	Trigonal planar
3	AX_2E	120°	SO_2	Trigonal planar	sp²	Bent
4	AX_4	109.5°	CH_4	Tetrahedral	sp³	Tetrahedral
4	AX_3E	109.5°	NH_3	Tetrahedral	sp³	Trigonal Pyramidal
4	AX_2E_2	109.5°	H_2O	Tetrahedral	sp³	Bent
5	AX_5	90°, 120°, 180°	PCl_5	Trigonal bipyramidal	sp³d	Trigonal Bipyramidal
5	AX_4E	90°, 120°, 180°	SF_4	Trigonal bipyramidal	sp³d	Seesaw
5	AX_3E_2	90°, 180°	ClF_4	Trigonal bipyramidal	sp³d	T – shape
5	AX_2E_3	180°	XeF_2	Trigonal bipyramidal	sp³d	Linear
6	AX_6	90°, 180°	SF_6	Octahedral	sp³d²	Octahedral
6	AX_5E	90°	BrF_5	Octahedral	sp³d²	Square Pyramidal
6	AX_4E_2	90°	XeF_4	Octahedral	sp³d²	Square Planar
6	AX_3E_3	90°, 180°		Octahedral	sp³d²	T – Shape
6	AX_2E_4	180°		Octahedral	sp³d²	Linear

Notes:

Chapter 10: Chemical Bonding II: Molecular Geometry and Bonding Theory

Solved Problem

Use valence bond theory to describe the bonding about an N atom in N_2F_4.

1. The Lewis electron-dot structure shows three bonds and one lone pair around each N atom
2. So that is four electron groups (accurately: AX_3E) on central atom
3. Therefore a tetrahedral arrangement
4. A tetrahedral arrangement has sp^3 hybrid orbitals

Explaining Multiple Bonds

- Consider CO_2 molecule. The Lewis structure is as follows:

  $\ddot{\underset{..}{O}}::C::\ddot{\underset{..}{O}}$ $\ddot{\underset{..}{O}}=C=\ddot{\underset{..}{O}}$

- It has carbon as central atom and two oxygen atoms as terminal atoms.
- The electronic geometry is AX_2
- The hybridization on carbon therefore is sp.

==================

- Now consider the HCN molecule. The Lewis structure is as follows:

  $H:C\equiv N:$

- The carbon is still central with H and N as terminal atoms.
- The electronic geometry is AX_2.
- The hybridization on carbon is still sp.

Solved Problem

What is the hybridization on the central atom in nitrate ion?

Answer:
- Lewis structure is
- Electronic geometry of N is AX$_3$
- Hybridization of a three electron group atom is sp^2

$$\left[\begin{array}{c} :\ddot{O}: \\ | \\ :\ddot{O} - N = \ddot{O}: \end{array} \right]^-$$

Notes:

Hybridization of Multiple Bonds

- Single bonds are formed by simple orbital overlap e.g. in H-H bond it is a s-s overlap.
- Some single bonds are hybridized (as discussed in previous slides). These are called sigma bonds.
- In a double bond there is a sigma and a pi bond.
- The pi bond is unhybrized orbital overlap of p orbitals.
- In a triple bond there is one sigma and two pi bonds.

Notes:

Chapter 10: Chemical Bonding II: Molecular Geometry and Bonding Theory

Ethylene – $CH_2=CH_2$

- Number of e- domains = 3
- Hybridization = sp^2 (shape = trigonal planar, bond angle = 120°)
- There are two central atoms; both carbon.
- Each carbon will mix 1 of s and 2 of p orbitals; 1 of p is left over and this forms the pi bond.

Double bond = 1 σ bond + 1 π bond

Notes:

Acetylene C_2H_2

$H-C\equiv C-H$

- Number of e- domains = 2
- Hybridization = sp (shape = linear, bond angle = 180°)
- There are two central atoms; both carbon.
- Each carbon will mix 1 of s and 1 of p orbitals; 2 of p orbitals are left over and this form two pi bond.

Triple bond = 1 σ bond + 2 π bonds

Notes:

Chapter 10: Chemical Bonding II: Molecular Geometry and Bonding Theory

Solved Problem

How many pi bonds and sigma bonds are in each of the following molecules? Describe the hybridization of each C atom.

(a) Cl—C(H)(H)—Cl, with C labeled sp^3

(b) H₂C=CH(Cl), both C labeled sp^2

(c) $H_3C-C=C-C\equiv C-H$, with hybridizations sp^3, sp^2, sp^2, sp, sp

(a) 4 sigma bonds

(b) 5 sigma bonds, 1 pi bond

(c) 10 sigma bonds, 3 pi bonds

Molecular Orbital Theory

- As atoms approach one another, their atomic orbitals overlap and form molecular orbitals.
- Two atomic orbitals combine to form two molecular orbitals. (4 AO will give 4 MO etc.) *(Half the of MO are bonding and half will be antibonding.)*
- How the orbitals are combining depends on energy and orientation. (Wavefuntions - + and − regions)
- Molecular orbitals concentrated in regions between nuclei (center of orbital) are called **bonding orbitals.** They are obtained by adding atomic orbitals (e.g. ψ+ and ψ+).
- Molecular orbitals having zero values in regions between nuclei (and are in other regions) are called **antibonding orbitals.** They are obtained by subtracting atomic orbitals (e.g. ψ+ and ψ-).

- (read more at: Chemwiki – UCDavis)

Chapter 10: Chemical Bonding II: Molecular Geometry and Bonding Theory

Molecular Orbital Theory

H + H → H_2
$1s^1$ $1s^1$ σ_{1s}^2

$1s$ $1s$
H atomic orbitals

H•   H—H   •H
antibonding orbital
σ_{1s}^*
σ_{1s}
bonding orbital

Notes:

Key Points

- Molecular geometry
  - VSEPR model
- Molecular geometry and polarity
- Valence bond theory
- Hybridization of atomic orbitals
  - s and p
  - s, p, and d
- Hybridization involving multiple bonds
- Molecular orbital theory
  - Bonding and antibonding orbitals

Notes:

Ch 10/ PowerPoint Study-2Hyrbridization   Name: _____

Answer these questions as you are watching the videos. They are due in class.
These questions are not just for you to answer but also to prepare you for the exam.
Make sure you understand what you are writing and not just copy from the text book. **Show all work.**

Fill out the chart using the guidelines you learned from the previous chapter.

Molecule/Ion	OF_2	H_2CO	NO_2^+	BF_3	SbF_5
Lewis Structure					
AXE formula					
Number of electron groups on central atom					
Electron Geometry					
Hybridization					
Bond angle(s)					
Shape of molecule					

Ch 10/Worksheet/Molecular Geometry   Name: _____

1. Write the structures between the following elements and name them

Elements	Formula	Name	Lewis Structure
K and O			
N and Cl			
Ca and N			

2. Use the models to make the models for the following molecules. Write the molecular geometry and molecular shape in the table below. Draw the 3D structure of the molecule.

	Lewis Structure	AXE Formula	Electronic Geometry	Molecular Shape	3D Structure
H_2O					
CO_2					
CH_3Cl					

Chapter 10: Chemical Bonding II: Molecular Geometry and Bonding Theory

	Lewis Structure	AXE Formula	Electronic Geometry	Molecular Shape	3D Structure
SO_2					
NO_3^-					

3. Complete the table below:

Compound	Lewis structure	VSEPR Formula (AXE)	Shape (draw out the shape)	Polarity of Molecule (yes/no)	Dipole Moment (show structure with arrow)
$OXeF_2$					
ClCN					
$CH_3NH_3^+$					

Ch 10/Worksheet/Bonding-Hybridization

Name: _____

1. Complete the following table.

Molecule	(1) <u>S</u>O2	(2) H<u>B</u>F2	(3) <u>Xe</u>F4	(4) <u>C</u>H₂Cl₂	(5) <u>N</u>F3
a) Lewis structure					
b) Electron groups on central atom.					
c) AXE formula					
d) Name of geometry on the central atom					
e) Bond angle associated with the geometry					
f) Name of shape of the molecule					
g) Hybridization					

2. Predict a) the bond angle, (b) the hybridization <u>around the indicated atoms (the atoms to which the arrows are drawn in the structures below).</u> Write your answers near the corresponding labels (1 to 5) in the drawings. (<u>Note</u>: the lone pairs on the F atoms are omitted.)

309